KS3
Success

Science

Practice Test Papers

Age 11-14

John
Beeby

Contents

Introduction

How to Use
the Practice Test Papers

About these Practice Test Papers

At the end of Key Stage 3, tests will be used by your teachers to determine your level of achievement in science.

In this book, you have three sets of test papers that will allow you to track your progress in Key Stage 3 Science. They will help you to identify your strengths and weaknesses in the subject. By sitting all three papers, you will be able to monitor your rate of progress.

The test papers will:
- test your knowledge and understanding of scientific facts and ideas, and how you use these to answer questions
- provide practice questions in all science topics
- help to familiarise you with the different question styles that appear in test papers
- highlight opportunities for further study and skills practice that will lead to improvement
- record results to track progress.

How to Use the Test Papers

The questions in these test papers have been written in the style that you will see in actual tests.

While you should try to complete the different sections in each set in the same week, you should complete sets A, B and C **at intervals** through Key Stage 3, or Year 9.

Make sure you leave a reasonable amount of time between each assessment – it is unrealistic to expect to see much improvement in just a few weeks. Spreading out the sets will mean you have an opportunity to develop and practise any areas you need to focus on. You will feel much more motivated if you wait for a while, because your progress will be much more obvious.

If you want to re-use the papers, write in pencil and then rub out the answers. However, don't repeat the set too soon or you will remember the answers and the results won't be a true reflection of your abilities.

Sets
ABC

KEY STAGE 3

Introduction

Science

Introduction

How to Prepare for the Tests

Revision: After covering the necessary science topics, read through your notes from school, or course notes. Perhaps use a revision guide to recap the key points. You can also add notes and diagrams to a mind map.

Equipment you will need:
- pen(s), pencil and rubber
- ruler
- protractor
- pair of compasses
- calculator
- a watch or clock to keep track of the pace at which you are answering questions.

When you feel that you're properly prepared, take the first set of test papers.

Taking the Tests

1. Each set of tests is made up of **two** test papers. Each paper is worth **75 marks**. You should spend **75 minutes (one and a quarter hours)** on **each** paper.
2. Choose a time to take the first paper when you can work through it in one go. Make sure you have an appropriate place to sit and take the test, and where you will be uninterrupted.
3. Answer **all** the questions in the test. If you are stuck on one question, move on and come back to it later. Tests often start with easier questions. These become more complex, and cover more than one topic, as you work through the test papers.
4. Read the questions **carefully**, so that you understand exactly what you need to do. Don't spend too long on any one question.
5. Write the answers in the spaces provided. The space provided for you to write your answer will also give you an indication of how detailed your answer needs to be. It will depend on the size of your writing, but if you need to use a lot more space, you're writing too much!
6. The number of marks allocated to each question is shown. This will tell you how many key points are needed in the answer.
7. Remember that marks may be awarded for key points or working out even if your final answer isn't correct, so **always show your working**, and keep it neat. It may be that if you get the answer wrong, you could still be awarded one mark for showing your working. Sometimes, the second mark for a calculation could be for the units of measurement, so make sure you include these.
8. Stay calm! Don't be fazed by questions. You may see some, for instance, that are based on topics that you haven't covered. When you do, it could be that the information you need to answer the question is provided in the question itself; or the question may be looking to see how you can apply your understanding to a new situation. Some questions will also test your understanding of how science works. You will not be able to revise for these, but practising doing this type of question will help.

How to Use the Mark Scheme

When you've sat the test, you, or a parent or guardian, should use the mark scheme to mark it. You could mark the test together. It's often helpful for you to discuss the answers with someone as you go through the mark scheme.

The answers and mark scheme will:
- give you an answer to the question **in full**. Any words shown in brackets aren't necessary to obtain the full marks, but should help your understanding of the question
- tell you where alternative answers are acceptable. If it's possible to use different words or terms in an answer, these will be separated by a forward slash, e.g. / . Sometimes when an answer isn't fully correct, certain alternatives may be acceptable
- provide Helpful Hints on answering particular questions.

When you've gone through the test paper, add up the marks to give you your total.

Tips for the Top

After sitting a test paper:
1. Try to analyse your performance. For questions that were incorrect, identify where you went wrong. Are there gaps in your knowledge and understanding? Were there topics where you were under-prepared? Have you misunderstood some of the science?
2. Pay attention to the Helpful Hints in the answers and mark scheme. These will give you revision tips, and important information about answering a question on a topic. They will also help you to avoid errors made by many students sitting tests.
3. Look for instructions in the question and the **command words**, as they will tell you exactly what kind of response is required, and the level of detail required in your answer.

For questions beginning:
- **Give, name** and **state** – you need to write down an answer that's short and concise. The answer could be from knowledge you have, or from information in the question.
- **Describe** – you need to write down a *more detailed* answer that gives the *key features* of something.
- **Explain** – you need to *give reasons* to answer how or why something happens.
- **Suggest** – you need to *apply* your knowledge to a new situation. This could be where you cannot give a conclusive answer or where you may not have enough evidence to draw a firm conclusion. Or, there may be more than one answer. Your suggestion could be based on your own scientific knowledge and understanding and/or information in the question.

When you've assessed your performance in the first test paper, do any additional work you need to. When you sit the second, and finally the third test, check to see how your performance is improving by comparing marks.

Test Paper 1

First name _____

Last name _____

Date _____

Instructions:

- The test is **75 minutes** long.

- Find a quiet place where you can sit down and complete the test paper undisturbed.

- You will need a pen, pencil, rubber and ruler. You may find a protractor and a calculator useful.

- The test starts with easier questions.

- Try to answer all of the questions.

- The number of marks available for each question is given in the margin.

- Show any rough working on this paper.

- Check your work carefully.

- Check how you have done using pages 101–112 of the Answers and Mark Scheme.

MAXIMUM MARK	75	ACTUAL MARK	

1. The diagram shows the structure of an atom.

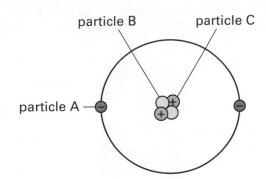

(a) (i) What is the name of subatomic particle A? Tick the correct box.

Electron ☐   Neutron ☐

Positron ☐   Proton ☐

(ii) What is the name of subatomic particle B? Tick the correct box.

Electron ☐   Neutron ☐

Positron ☐   Proton ☐

(iii) What is the name of subatomic particle C? Tick the correct box.

Electron ☐   Neutron ☐

Positron ☐   Proton ☐

(b) Complete the table below.

Particle	Relative mass
Electron	
Neutron	

2. We all look different. We show variation.

Human characteristics are a combination of genetic and environmental variation.

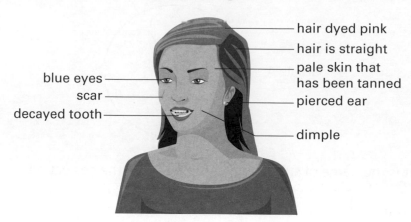

blue eyes
scar
decayed tooth

hair dyed pink
hair is straight
pale skin that has been tanned
pierced ear
dimple

(a) From the picture of Jade, above:

(i) Give **two** characteristics that are controlled by her genes.

_____

_____

(ii) Give **two** characteristics that are the result of her environment.

_____

_____

(b) Give **two** characteristics that are controlled by genes but are affected by the environment.

_____

_____

(c) A study investigated certain characteristics of the men in a town.

The table shows the percentage of men with different blood groups.

Blood group	% of men with blood group
A	42
B	10
AB	4
O	44

1 mark

1 mark

1 mark

(i) Label and plot a graph of the results.

2 marks

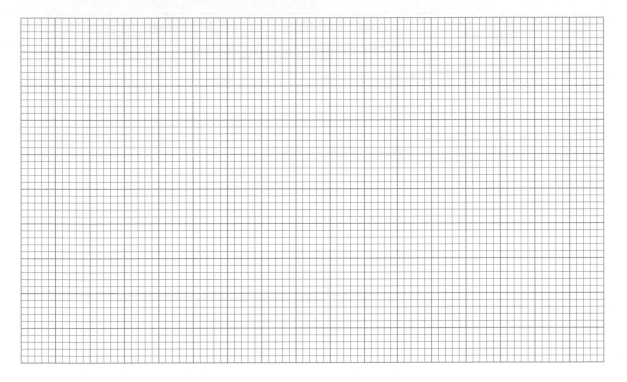

(ii) Here is a graph that shows the number of men of different height ranges in the town.

2 marks

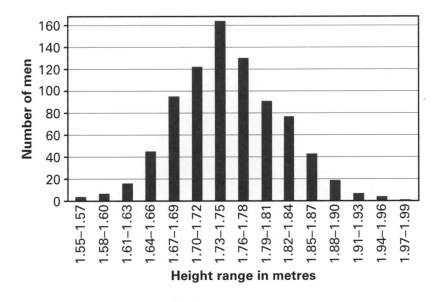

Use information from both graphs.

What type of variation is shown by:

Blood groups? _____

Height? _____

SUBTOTAL

3. Complete the table below on energy transfers.

Process	Type of energy at the start	Final energy forms	
		Wanted forms	Wasted forms
Switching on a torch		light	
Lighting a Bunsen burner			light
Releasing a stretched spring	elastic potential		
Releasing a toy car at the top of a ramp		kinetic	

4. These diagrams show the male and female reproductive systems of humans.

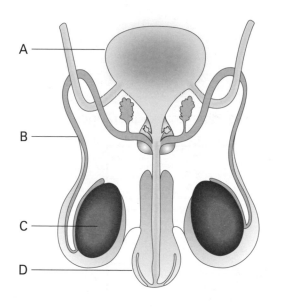 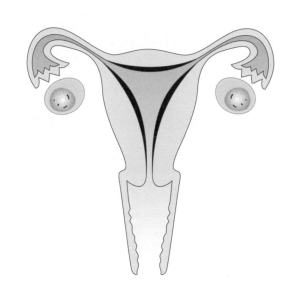

(a) In the male reproductive system:

(i) Which letter shows the testis? Tick the correct box.

A ☐   B ☐

C ☐   D ☐

(ii) Which letter shows the sperm duct? Tick the correct box.

A ☐   B ☐

C ☐   D ☐

(b) On the diagram of the female reproductive system, label the:

    (i) ovary

1 mark

    (ii) vagina

1 mark

(c) In which organs are the sex cells produced?

1 mark

_____

(d) Where does fertilisation occur?

1 mark

_____

5. Iron is extracted from its ore in a blast furnace.

(a) The ore used in the blast furnace is haematite. It has the formula Fe_2O_3.

    (i) How many atoms are present in the formula of haematite?

1 mark

_____

    (ii) The word equation for one of the chemical reactions in the blast furnace is:

1 mark

        iron oxide   +   carbon   →   iron   +   carbon dioxide

    What is happening to the carbon in this reaction?

_____

(b) One ore of aluminium is bauxite, which is aluminium oxide. Explain why aluminium cannot be extracted from aluminium oxide using carbon.

2 marks

_____

_____

_____

(c) In nature, the elements gold and silver are usually found as the metals themselves and not compounds. Sodium is never found as sodium, but always as sodium compounds. Explain why.

2 marks

_____

_____

SUBTOTAL

6. Kayleigh is a lighting technician in a theatre.

She puts filters over the theatre lights.

stage lighting white light	stage lighting white light	stage lighting white light

blue filter   green filter   red filter

coloured light A   coloured light B   coloured light C

(a) What colour is:

Coloured light A? _____

Coloured light B? _____

Coloured light C? _____

(b) The actors' costumes absorb and reflect different colours.

(i) What colour will a red shirt appear under red light? Tick the correct box.

Black	Blue	Red	White	Yellow

(ii) What colour will a blue dress appear under red light? Tick the correct box.

Black	Blue	Red	White	Yellow

(iii) What colour will a yellow tie appear under red light? Tick the correct box.

Black	Blue	Red	White	Yellow

7. This is a diagram of the carbon cycle.

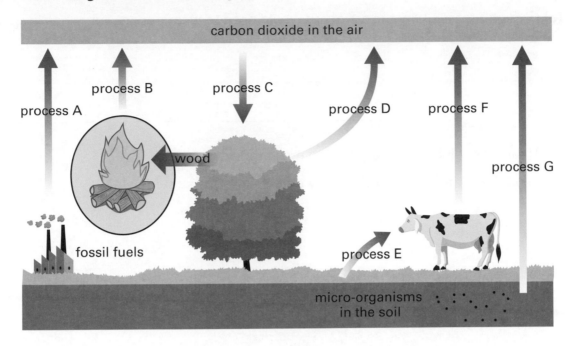

(a) Give the letters of the processes that release carbon dioxide into the air?

_____

_____

2 marks

(b) Give the letters of the processes where oxidation reactions?

_____

_____

2 marks

(c) Which process is photosynthesis?

_____

1 mark

(d) What is the name of processes A and B?

_____

1 mark

SUBTOTAL

2 marks

8. An astronaut is testing a new space suit.

(a) Give **one** reason why the astronaut would need to wear a space suit when in space. Explain your answer.

____

____

2 marks

(b) The space suit has four jet packs to move the astronaut.

In one test, the forces produced by the jet packs are shown below.

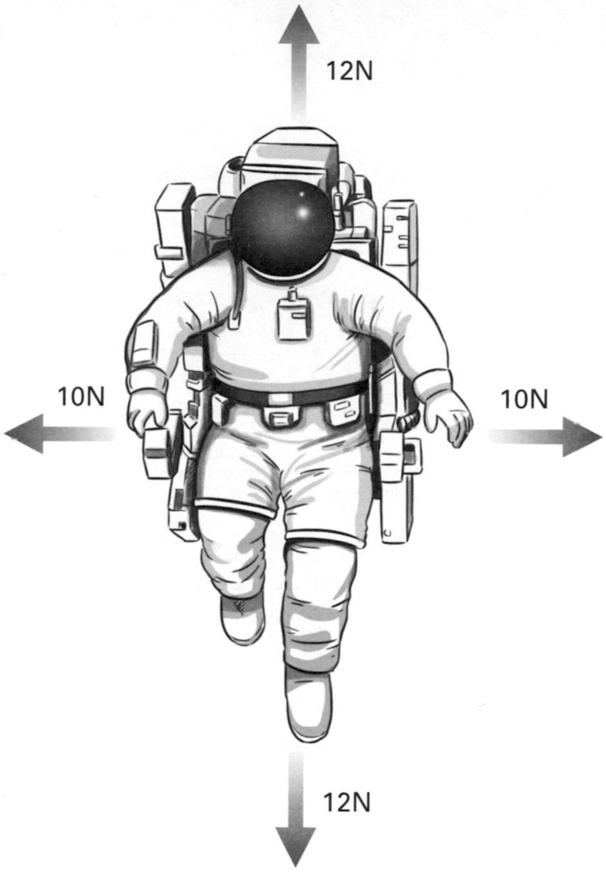

In which direction will the astronaut move? Explain your answer.

____

____

____

1 mark

(c) The astronaut is on the Moon.

He has a mass of 100kg.

The gravitational field strength on the Moon is 1.6N/kg.

How much will he weigh?

____________________ N

9. This picture shows an important biological molecule.

(a) What is the name of this biological molecule?

_____

1 mark

(b) Name the **four** scientists whose work led to finding out the structure of this molecule.

_____

_____

_____

2 marks

(c) Where is this molecule found in our cells?

_____

1 mark

(d) Explain why this molecule is so important in forensic science investigations.

_____

_____

_____

_____

_____

2 marks

SUBTOTAL

10. The current Periodic Table is shown below.

1	2											3	4	5	6	7	0
						1 H hydrogen 1											4 He helium 2
7 Li lithium 3	9 Be beryllium 4											11 B boron 5	12 C carbon 6	14 N nitrogen 7	16 O oxygen 8	19 F fluorine 9	20 Ne neon 10
23 Na sodium 11	24 Mg magnesium 12											27 Al aluminium 13	28 Si silicon 14	31 P phosphorus 15	32 S sulfur 16	35.5 Cl chlorine 17	40 Ar argon 18
39 K potassium 19	40 Ca calcium 20	45 Sc scandium 21	48 Ti titanium 22	51 V vanadium 23	52 Cr chromium 24	55 Mn manganese 25	56 Fe iron 26	59 Co cobalt 27	59 Ni nickel 28	63.5 Cu copper 29	65 Zn zinc 30	70 Ga gallium 31	73 Ge germanium 32	75 As arsenic 33	79 Se selenium 34	80 Br bromine 35	84 Kr krypton 36
85 Rb rubidium 37	88 Sr strontium 38	89 Y yttrium 39	91 Zr zirconium 40	93 Nb niobium 41	96 Mo molybdenum 42	98 Tc technetium 43	101 Ru ruthenium 44	103 Rh rhodium 45	106 Pd palladium 46	108 Ag silver 47	112 Cd cadmium 48	115 In indium 49	119 Sn tin 50	122 Sb antimony 51	128 Te tellurium 52	127 I iodine 53	131 Xe xenon 54
133 Cs caesium 55	137 Ba barium 56	139 La* lanthanum 57	178 Hf hafnium 72	181 Ta tantalum 73	184 W tungsten 74	186 Re rhenium 75	190 Os osmium 76	192 Ir iridium 77	195 Pt platinum 78	197 Au gold 79	201 Hg mercury 80	204 Tl thallium 81	207 Pb lead 82	209 Bi bismuth 83	209 Po polonium 84	210 At astatine 85	222 Rn radon 86
223 Fr francium 87	226 Ra radium 88	227 Ac* actinium 89	261 Rf rutherfordium 104	262 Db dubnium 105	266 Sg seaborgium 106	264 Bh bohrium 107	277 Hs hassium 108	268 Mt meitnerium 109	271 Ds darmstadtium 110	272 Rg roentgenium 111							

Key:
relative atomic mass
atomic symbol
name
atomic (proton) number

Elements with atomic numbers 112–116 have been reported but not fully authenticated

(a) Which scientist published the first true Periodic Table?

_____

1 mark

(b) What principles did he use to arrange the elements in the Periodic Table?

_____

_____

2 marks

(c) How would scientists describe the elements that are bordered on the far right of the table?

_____

1 mark

(d) What term would scientists use to describe one of the columns of the Periodic Table?

_____

1 mark

11. Zainab is a food analyst. She is investigating the quality of a sample of octadecanoic acid.

She puts a sample of solid octadecanoic acid into a test tube and heats it.

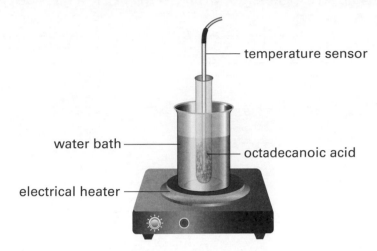

(a) Zainab uses a datalogger to measure the temperature of her sample every two minutes. Here are her results:

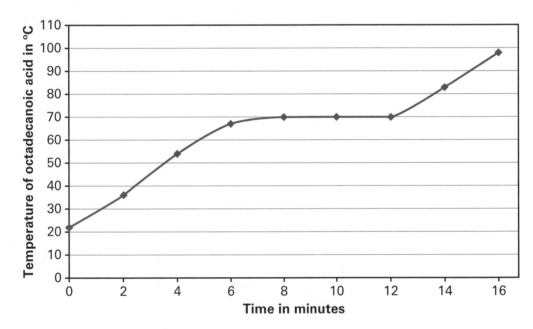

Explain what is happening between 8 and 12 minutes.

_____

_____

_____

3 marks

(b) From the graph, Zainab can tell that the octadecanoic acid is pure. Explain how she knows.

_____

_____

_____

2 marks

SUBTOTAL

12. Scarlett is learning about how sound travels.

Her teacher sets up an experiment with an electric bell in a glass bell jar.

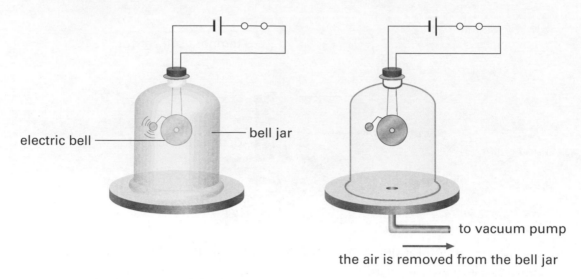

electric bell

bell jar

to vacuum pump

the air is removed from the bell jar

(a) Explain why she could not hear the bell when all the air was removed from the bell jar.

_____

_____

(b) The table below gives the speed of sound as it travels through different media.

Medium	Speed of sound in m/s
Air	330
Alcohol	1160
Brick	4200
Steel	6100
Water, distilled	1490
Wood (hardwood)	4000

(i) Draw a bar chart to display the data.

3 marks

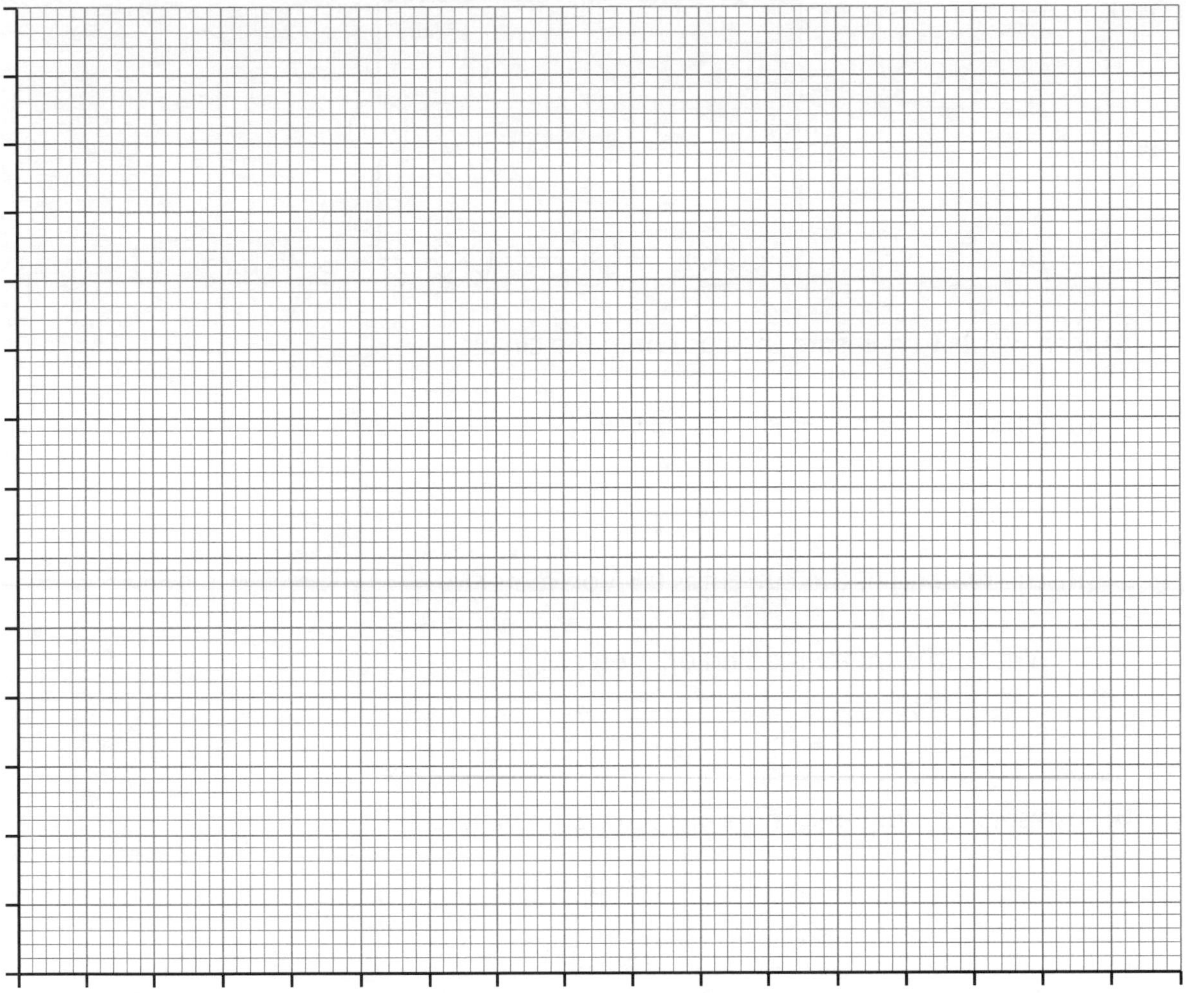

(ii) Compare the speed of sound in the listed solids, liquids and the gas. Suggest an explanation as to the differences.

2 marks

_____

_____

_____

(c) Explain how sound is used by ships to work out the depth of water.

3 marks

_____

_____

_____

_____

SUBTOTAL

13. Josh sets up the electrical circuit shown below.

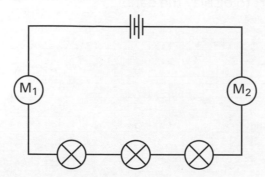

(a) Josh wants to measure the current in the circuit.

(i) What type of meter should he use?

_____

(ii) The reading of the current is 0.5A on meter M_1.

What is the meter reading on meter M_2?

_____

(b) Josh sets up a different type of circuit, shown here.

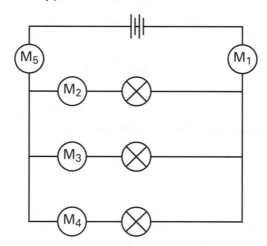

The circuit has two cells and three light bulbs – the same as the previous circuit.

(i) What is the name of this type of circuit?

_____

(ii) The reading on meter M_1 is 0.6A, and on meters M_2 and M_3 it is 0.2A.

What is the reading on:

Meter M_4? _____

Meter M_5? _____

14. A toxic insecticide is sprayed on farmland to control insect pests. Some of the insecticide ends up in the sea.

(a) Here is a food web diagram of some of the organisms affected.

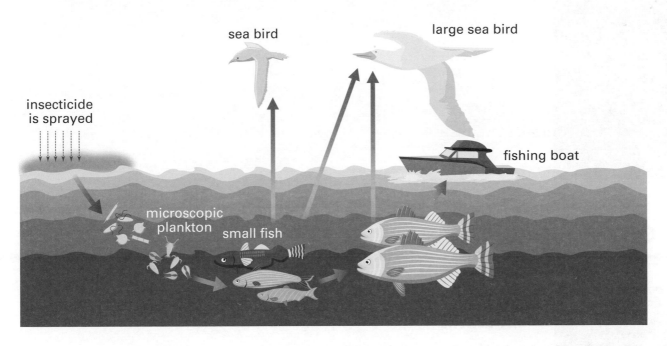

(i) Suggest why the sea birds start to die but the microscopic plankton and the fish are not killed.

3 marks

_____

_____

_____

(ii) Explain how, after some time, the insecticide begins to appear in humans.

1 mark

_____

SUBTOTAL

Test Paper 2

First name _____

Last name _____

Date _____

Instructions:

• The test is **75 minutes** long.

• Find a quiet place where you can sit down and complete the test paper undisturbed.

• You will need a pen, pencil, rubber and ruler. You may find a protractor and a calculator useful.

• The test starts with easier questions.

• Try to answer all of the questions.

• The number of marks available for each question is given in the margin.

• Show any rough working on this paper.

• Check your work carefully.

• Check how you have done using pages 101–112 of the Answers and Mark Scheme.

MAXIMUM MARK	75		ACTUAL MARK	

1. Jasmine is eating a cereal bar. She looks at the nutritional information on the pack.

Nutritional information		
Typical values	**Per 100 g**	**Per bar**
Energy kJ kCal	1900 452	760 181
Protein	6.0	2.4
Carbohydrate	52.0	20.8
Fat (lipids)	24.5	9.8
Fibre	4.7	1.9

(a) Why does the information refer to 'Typical values' per 100 g and per bar?

_____

_____

1 mark

(b) Calculate the mass of the cereal bar that Jasmine is eating. Show your working out.

_____ **g**

2 marks

(c) Write down one function of each of the following:

Protein

_____

Carbohydrate

_____

Fat (lipids)

_____

3 marks

SUBTOTAL

(d) Which nutrients essential for health are missing from the nutritional information?

_____

_____

(e) Jasmine reads in a magazine that teenage girls should eat no more than 20g of saturated fat every day.

Each cereal bar contains 4g.

If all her saturated fat came from cereal bars, what would be the maximum number of bars she should eat in a day?

_____

2. The photographs show some ways of generating electricity.

Nuclear

Coal

Solar

Wind

(a) Which of these ways of producing electricity:

(i) depends on fossil fuel?

_____

(ii) are based on renewable energy sources?

_____

(iii) produce no carbon dioxide during electricity production?

_____

(iv) can be used at home to produce electricity?

_____

(b) The diagram shows how hydroelectric power is produced.

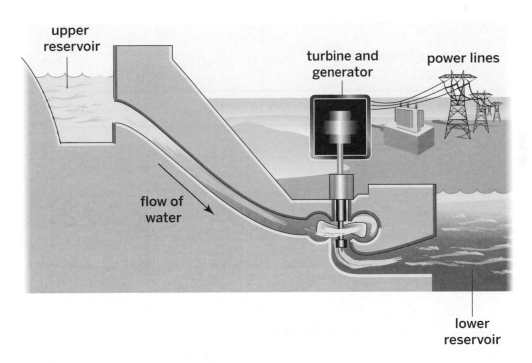

Write down the energy transfers in a hydroelectric power plant.

_____

_____

_____

3. The graph below shows the physical states of carbon dioxide at different temperatures and pressures.

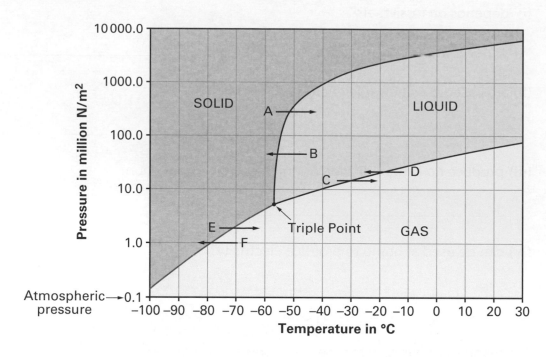

(a) At which of the following temperatures and pressures is carbon dioxide a liquid? Use the graph to help you.

Tick the correct box in the table below.

Temperature in °C	Pressure in million N/m²	Liquid?
30	10.0	
20	1000.0	
0	10 000.0	
−60	1.0	

(b) At what temperature does carbon dioxide boil at a pressure of 10 million N/m²?

_____

(c) What happens to the melting point of carbon dioxide as the pressure is increased?

_____

(d) The arrows on the graph show state changes of carbon dioxide.

(i) Which one of the arrows shows carbon dioxide freezing? Tick the correct box.

1 mark

A   B   C   D   E

☐   ☐   ☐   ☐   ☐

(ii) Which one of the arrows shows carbon dioxide boiling? Tick the correct box.

1 mark

A   B   C   D   E

☐   ☐   ☐   ☐   ☐

(iii) Sublimation is a process where a solid changes to a gas without going through a liquid state.

1 mark

Which one of the arrows shows sublimation? Tick the correct box.

A   B   C   D   E

☐   ☐   ☐   ☐   ☐

(e) A point called the Triple Point is marked on the graph.

(i) Suggest what the Triple Point is.

1 mark

_____

(ii) At what temperature does the Triple Point occur?

1 mark

_____

SUBTOTAL

☐

4. The diagrams below show the structure of the tissues that form the lining of the mouth and the trachea (windpipe).

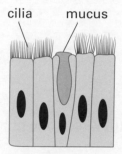

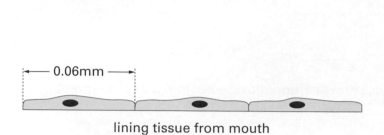

lining tissue from mouth   lining tissue from trachea

(a) Give two differences in the structure of the tissue lining the mouth and the tissue lining the trachea.

_____

_____

_____

_____

(b) The actual width of the cell shown is 0.06mm.

Calculate the magnification of the diagram. Show your working.

$$\text{Magnification} = \frac{\text{size on diagram}}{\text{actual size}}$$

Magnification $= x$ _____

5. Megan sits on a see-saw.

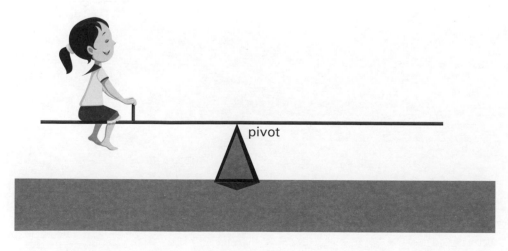

(a) What type of turning moment does Megan produce?

1 mark

____

(b) Megan weighs 350N. She sits 2m away from the pivot of the see-saw.
Calculate the turning moment produced by Megan. Show your working-out.

2 marks

____ Nm

(c) Thomas joins Megan on the see-saw.

2 marks

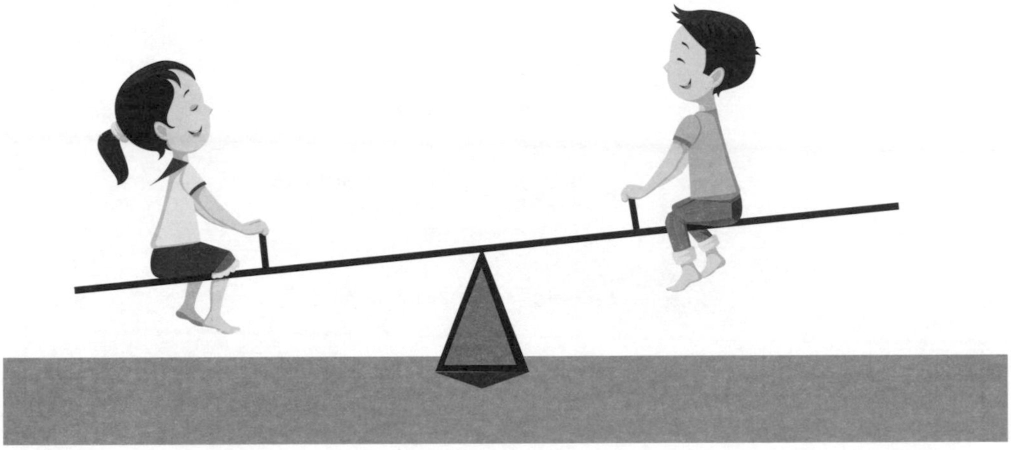

Thomas moves about until the see-saw is balanced. He is now sitting 1.75m from the pivot. Calculate Thomas's weight.

____ N

(d) Megan's dog sits on the see-saw with her.
The dog weighs 20N and sits 1.5m away from the pivot.
Calculate the turning moment that is now on the left-hand side of the see-saw.

2 marks

____ Nm

SUBTOTAL

6. James is investigating the chemical reaction between zinc and copper sulfate.

He measures out some copper sulfate solution into a beaker and records its temperature.

He then weighs out some zinc powder and adds it to the copper sulfate solution.

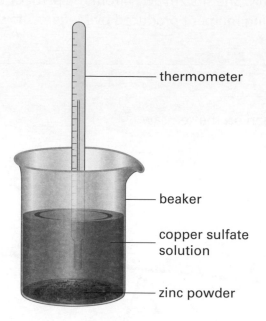

James records the temperature rise.

He then repeats the investigation with different masses of zinc powder.

Here are his results:

Mass of zinc powder in g	Temperature rise in °C
0	0.0
1	4.5
2	10.0
3	14.5
4	20.0
5	25.0
6	29.5

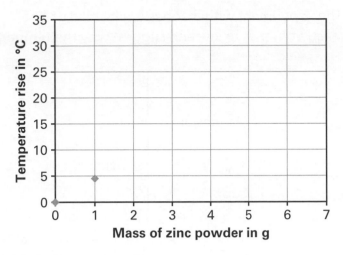

Temperature rise in °C

Mass of zinc powder in g

(a) (i) Plot a graph of James's results.
The first two points have been done for you.

(ii) Draw the best line through the points.

(b) Jess carries out a similar investigation.
She uses a polystyrene coffee cup instead of a beaker.
She also covers the cup with a lid.
Whose experiment will produce more reliable results?
Explain why.

_____

_____

_____

_____

_____

(c) A chemical reaction has taken place between the zinc and copper sulfate.
Give a word equation for this chemical reaction.

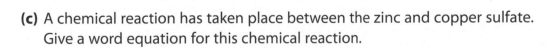

_____   +   _____   ➔   _____   +   _____

(d) Jess repeats the experiment with magnesium instead of zinc.
Look at the reactivity series below and suggest why the temperature rise is greater.

Sodium

Calcium

Magnesium

Aluminium   Decreasing

Zinc   reactivity

Copper

Silver

_____

_____

_____

7.   Agnieszka is brewing some beer.

(a) Which micro-organism is used to brew beer?

_____

(b) This organism can respire with or without oxygen.

(i) Write a full symbol equation for **aerobic** respiration. The formula for glucose is $C_6H_{12}O_6$.

_____ + _____ $\rightarrow$ _____ + _____ (+ ENERGY)

(ii) Write a word equation for **anaerobic** respiration.

_____ $\rightarrow$ _____ + _____ (+ ENERGY)

(c) Agnieszka took some measurements to find the density of her beer.

Use the formula:

$$\text{density in g per cm}^3 = \frac{\text{mass in g}}{\text{volume in cm}^3}$$

The mass of her sample of beer = 525g

The volume of her sample of beer = 500cm^3

_____ g per cm^3

8. The Earth rotates on its axis and orbits the Sun.

This diagram shows the Earth's orbit around the Sun.

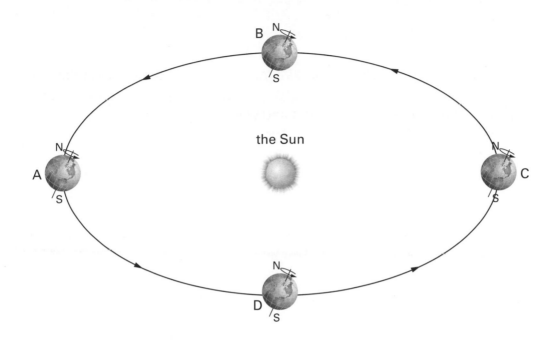

(a) (i) Which position shows the Earth when it is winter in the Northern Hemisphere?

1 mark

Tick the correct box.

Position A ☐   Position B ☐

Position C ☐   Position D ☐

(ii) Explain why, when it is winter in the Northern Hemisphere, it is summer in the Southern Hemisphere.

2 marks

_____

_____

_____

_____

(b) (i) How long does the Earth take to orbit around the Sun?

1 mark

_____

SUBTOTAL

(ii) How long does it take the Earth to do a complete rotation on its axis?

_____

9. Alex is making some fertiliser in the lab.

He reacts nitric acid with the alkali potassium hydroxide in the correct quantities.

(a) Acids react with alkalis to form a salt and water.
Summarise this reaction using a word equation.

_____ + _____ → _____ + _____

(b) (i) How will the pH of the mixture change as the alkali is added to the acid?

_____

_____

_____

(ii) What could he use to check that he has added the correct volumes of acid and alkali?

_____

(c) The formula of potassium hydroxide is KOH.
The formula of nitric acid is HNO_3.
Write a symbol equation for the reaction between potassium hydroxide and nitric acid.

_____ + _____ → _____ + _____

(d) The reaction produces a solution of the fertiliser.
How can Alex produce solid fertiliser?

_____

_____

_____

10. In her PE lesson, Olivia breathes into a spirometer to measure the volume of air in her lungs.

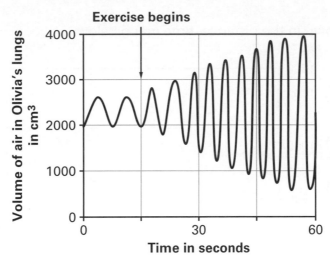

(a) How long after breathing into the spirometer does Olivia begin her exercise?

1 mark

_____

(b) As she begins to exercise, her breathing changes.

    (i) Write down two ways in which Olivia's breathing changes.

2 marks

    1 _____

    _____

    2 _____

    _____

    (ii) Explain why these changes in her breathing occur.

2 marks

    _____

    _____

    _____

    _____

    _____

SUBTOTAL

11. William gets on a bus and makes a journey of 1500 metres.

His journey is shown in the distance–time graph below.

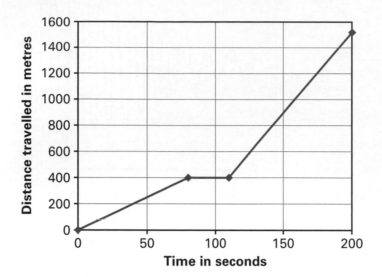

(a) How long does William's bus journey take?

_____

(b) Calculate the average speed of the bus over the journey.

_____

(c) When did the bus stop to pick up passengers?

_____

(d) (i) During which period was the bus travelling fastest?

_____

(ii) Calculate the speed of the bus during this part of the journey.

_____

1 mark

1 mark

1 mark

1 mark

3 marks

Test Paper 1

First name _____

Last name _____

Date _____

Instructions:

- The test is **75 minutes** long.

- Find a quiet place where you can sit down and complete the test paper undisturbed.

- You will need a pen, pencil, rubber and ruler. You may find a protractor and a calculator useful.

- The test starts with easier questions.

- Try to answer all of the questions.

- The number of marks available for each question is given in the margin.

- Show any rough working on this paper.

- Check your work carefully.

- Check how you have done using pages 101–112 of the Answers and Mark Scheme.

MAXIMUM MARK	75		ACTUAL MARK	

1. Huw put some copper wire into a test tube of silver nitrate solution and left the test tube undisturbed for one day.

Here are his results.

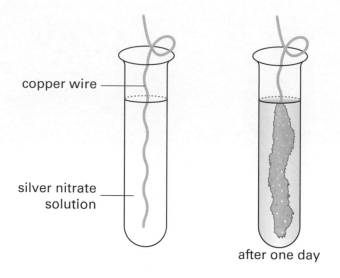

copper wire

silver nitrate solution

after one day

(a) A chemical reaction has taken place.

(i) What is the name of this type of chemical reaction?

_____

(ii) Write a word equation for the chemical reaction that has taken place.

_____ + _____ → _____ + _____

_____ _____ _____ _____

(b) Huw repeats the experiment with different metals.

He adds different metal nitrate solutions to the metals.

His results are shown in the table below.

Solution	Metal			
	Copper	**Iron**	**Lead**	**Tin**
Copper nitrate	no reaction	reaction	reaction	reaction
Iron nitrate	no reaction	no reaction	no reaction	no reaction
Lead nitrate	no reaction	reaction	no reaction	reaction
Tin nitrate	no reaction	reaction	no reaction	no reaction

(i) Use these results to put the metals in order of their reactivity.

Most reactive

_____

_____

_____

_____

Least reactive

(ii) Where does silver fit into this reactivity series? Explain why.

2 marks

_____

_____

_____

2. This question is about lenses and mirrors.

(a) (i) Complete the sentences below by choosing the correct words from the box.

3 marks

| reflection | refraction | slows down | speeds up |

Light entering the lens _____ and changes direction. This is called

_____ .

As light leaves the lens, moving from glass to air, it _____ ,
so light rays are bent the other way.

(ii) Give **two** examples of where lenses are used.

1 mark

1 _____

2 _____

(b) The following diagram shows a mirror.

incoming ray

angle *i*

Which **two** statements about mirrors are correct?

Tick the correct boxes.

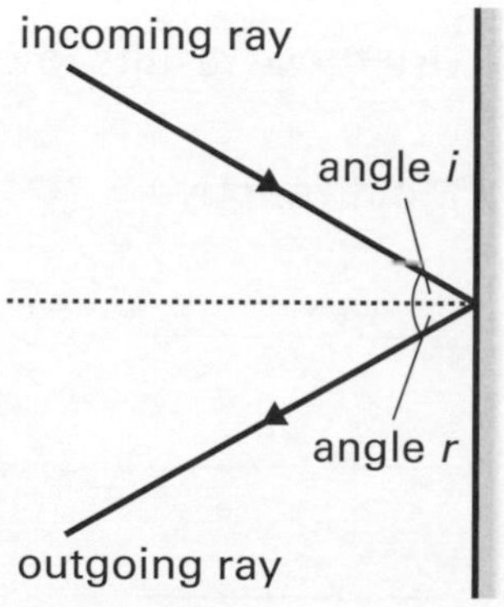

angle *r*

outgoing ray

Angle *i* is larger than angle *r* ☐   A mirror image is reversed, left to right ☐

Angle *i* equals angle *r* ☐   Light is refracted by a mirror ☐

3. This question is about cells.

(a) Complete the sentences below about plant cells by choosing the correct words from the box.

cell wall	chloroplast	cytoplasm	nucleus

The _____ forms the outer layer of the cell that supports the organism.

The _____ absorbs light for photosynthesis. The _____ controls the activities of each cell.

(b) The diagrams below show some different types of cell.

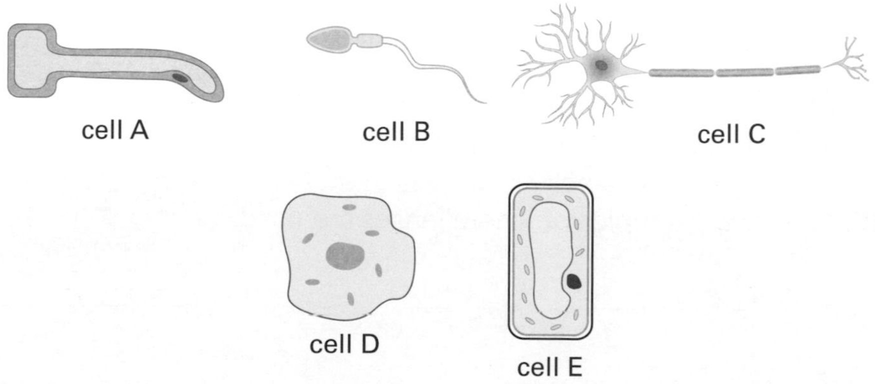

cell A   cell B   cell C

cell D

cell E

(i) Give the letters of the two types of plant cell.

_____ and _____

40

(ii) Which cell is involved in reproduction?

_____

(iii) Which cell communicates with other parts of the organism?

_____

(iv) Name the structures close to the nucleus in cell D.

_____

(v) What process is involved in the movement of gases in and out of cell E?

_____

4.   Lizzie is carrying out an experiment with a spring.

She adds different masses and measures the increase in length of the spring.

(a) Lizzie measures the force produced by the masses in units with the symbol N.

(i) What does this symbol, N, stand for? Tick the correct box.

Neutral ☐   Neutron ☐

Newton ☐   Nitrogen ☐

(ii) What type of force are the masses producing on the spring? Tick the correct box.

Compression ☐   Push ☐

Squash ☐   Stretch ☐

(b) Lizzie's results are shown below.

Force in N	Increase in length of spring in mm
0.0	0
0.5	20
1.0	40
1.5	60
2.0	80
2.5	100

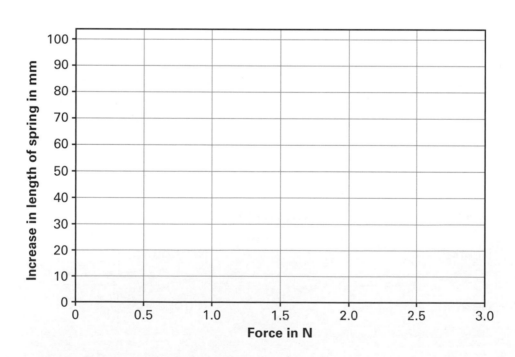

(i) Plot a line graph in the grid above.

(ii) Draw the line of best fit through the points.

(iii) Describe the relationship between the force and increase in length of the spring.

_____

_____

_____

5. This diagram shows the structure of the human lungs.

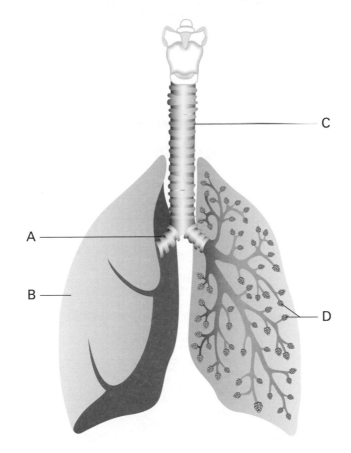

(a) (i) Which letter on the diagram labels the trachea? Tick the correct box.

1 mark

A ☐ B ☐

C ☐ D ☐

(ii) Which letter on the diagram labels the bronchus? Tick the correct box.

1 mark

A ☐ B ☐

C ☐ D ☐

(b) Jodie's teacher sets up a model of the lungs in class.

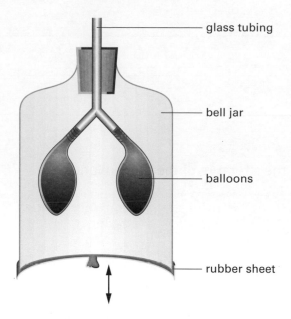

glass tubing

bell jar

balloons

rubber sheet

(i) Which part of the model represents the lungs?

_____

(ii) What does the rubber sheet represent?

_____

(iii) Explain how changes in the chest result in air being breathed in.

_____

_____

_____

_____

_____

6. The diagrams below show the arrangement of particles in a solid, liquid and gas.

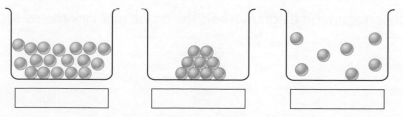

(a) Label the diagrams as solid, liquid or gas.

(b) When a liquid is heated, it evaporates. Describe what happens in this process.

2 marks

_____

_____

_____

_____

(c) Explain why it is easy to compress a gas, but very difficult to compress a liquid.

1 mark

_____

_____

_____

_____

7.   Larry is a sound engineer in a recording studio.

(a) He uses a digital oscilloscope to look at sound vibrations from his music.

Some of these are shown below.

A 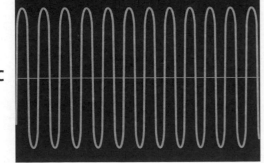 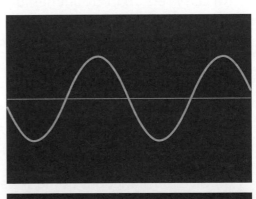 B

C 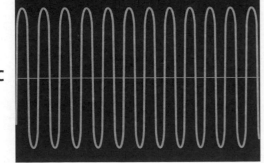 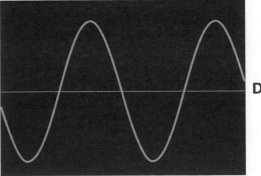 D

SUBTOTAL

1 mark

(a) (i) Which sound is quietest?

Tick the correct box.

A ☐   B ☐

C ☐   D ☐

1 mark

(ii) Which sounds have the lowest frequency?

Tick the **two** correct boxes.

A ☐   B ☐

C ☐   D ☐

1 mark

(iii) In which unit is frequency measured? Tick the correct box.

Amplitude ☐   Decibels ☐

Hertz ☐   Lux ☐

2 marks

(b) Larry sets up microphones for a recording session.

What energy transfer takes place in a microphone?

_____

8. Humans have an internal skeleton.

2 marks

(a) Give **four** functions of the skeleton.

1 _____

2 _____

3 _____

4 _____

(b) Jason is weight training. He wants to measure the force produced by his biceps muscle.

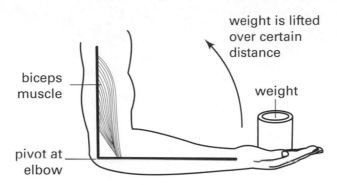

weight is lifted
over certain
distance

biceps
muscle

weight

pivot at
elbow

Give three measurements that he needs to make, or values he has to record, to be able to do the calculation.

1 _____

2 _____

3 _____

9.   The table below gives the names and formulae of some acids.

Name of acid	Carbonic	Ethanoic	Hydrochloric	Hydrocyanic	Nitric	Sulfuric
Chemical formula	H_2CO_3	CH_3COOH	HCl	HCN	HNO_3	H_2SO_4

(a) Which element is found in all of the acids?

_____

(b) The chart shows the pH scale and the colours obtained with Universal indicator.

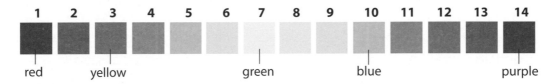

1   2   3   4   5   6   7   8   9   10   11   12   13   14

red   yellow   green   blue   purple

(i) What is the pH range of Universal indicator?

_____

(ii) What colour would sulfuric acid give with Universal indicator?

_____

(iii) What colour would pure water give with Universal indicator?

_____

3 marks

1 mark

1 mark

1 mark

1 mark

SUBTOTAL

(c) Owen decides to use a pH meter to find a liquid's pH.

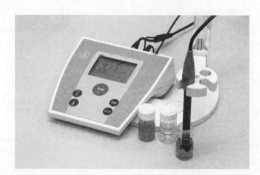

Give one advantage and one disadvantage of using a pH meter to find the pH of a liquid.

Advantage: _____

_____

Disadvantage: _____

_____

10. Jasmine is looking at the energy requirements of different females.

Female	Energy requirement in kJ per day
8 year old, active	8000
15 year old, active	12 000
Woman, office worker	10 000
Woman, breast-feeding	18 000

(a) Label and plot a bar chart of the energy requirements of different females.

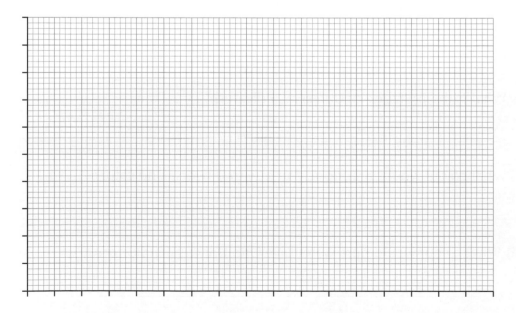

(b) Explain the need for extra energy for the mother who is breast-feeding.

_____

(c) Suggest how and why the energy requirements of a 15-year-old male would be different from a 15-year-old female.

_____

_____

11. Scientists researching the way our atmosphere has changed have published the following data:

Gas	Today	Number of millions of years ago			
		1000	**2000**	**3000**	**4000**
Carbon dioxide %	0.04	1	3	10	20
Nitrogen %	78	77	72	54	35
Oxygen %	21	10	1	0	0
Other gases %	0.96		24	36	45

(a) Complete the table by calculating the percentage of other gases in the air 1000 million years ago.

(b) Draw a pie chart to show the composition of the atmosphere 4000 million years ago.

(c) Animals are thought to have appeared on Earth around 750 million years ago. Suggest why they did not appear before then.

_____

12. Jenny lives in an old house.
The drawing shows what percentage of heat energy is lost through different parts of it.

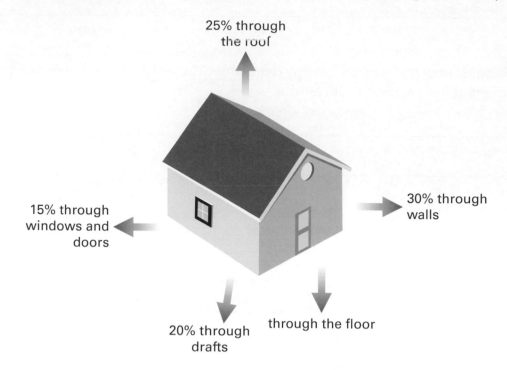

25% through the roof

15% through windows and doors

30% through walls

20% through drafts

through the floor

(a) What percentage of heat is lost through the floor? (Assume that heat is lost in no other way except for those shown.)

_____

(b) Jenny looks at the cost of some methods to reduce heat loss from her home. These are shown in the table below.

Method of reducing heat loss	Cost in £
Cavity wall insulation	350
Double-glazed windows	1500
Draft excluders around doors and windows	100
Loft insulation – 100mm	300
Loft insulation – 270mm	450

(i) Jenny has a limited budget. Explain why she should not spend her money on double-glazing.

_____

_____

(ii) Jenny decides to have her loft insulated.

She is told by her energy company that:

- if she has 100mm-thick insulation, she will save £25 per year on her energy bills
- if she has 270-mm thick insulation, she will save £150 per year.

With savings she would make, calculate how long it will take her to cover her costs if she has:

100mm-thick insulation _____

_____

270mm-thick insulation _____

_____

_____

(c) Complete the sentences below by choosing the correct words from the box.

conduction	convection	radiation

Heat energy gets transferred through the walls, roof, windows and floor of our homes

by _____

Cold air entering the house as a draft can take heat energy into the loft by

_____

13. The table below shows the feeding relationships of organisms in a UK woodland.

Organism	Feeds on
Robin	Caterpillars
	Earthworms
	Woodlice
Caterpillar	Plants
Earthworm	Plants
Shrew	Caterpillars
	Earthworms
	Woodlice
Hawk	Robins
	Shrews
Woodlouse	Plants

(a) What type of organism is the producer in the food web?

_____

(b) Draw a food web to show the feeding relationships.

Test Paper 2

First name _____

Last name _____

Date _____

Instructions:

- The test is **75 minutes** long.

- Find a quiet place where you can sit down and complete the test paper undisturbed.

- You will need a pen, pencil, rubber and ruler. You may find a protractor and a calculator useful.

- The test starts with easier questions.

- Try to answer all of the questions.

- The number of marks available for each question is given in the margin.

- Show any rough working on this paper.

- Check your work carefully.

- Check how you have done using pages 101–112 of the Answers and Mark Scheme.

MAXIMUM MARK	75		ACTUAL MARK	

1. Josh is investigating the electrical resistance of a length of wire.

He sets up the circuit below.

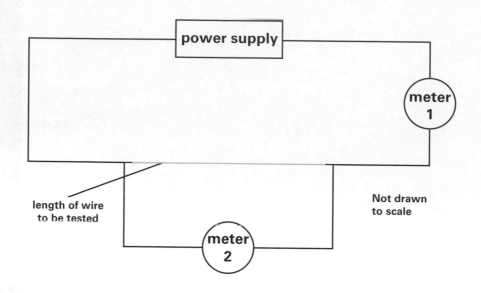

(a) He measures potential difference and the current.

Which does he measure on:

Meter 1? _____   Meter 2? _____

(b) He records a potential difference of 1.10V and a current of 0.65A.

Calculate the resistance of the wire using the formula:

$$\text{Resistance in ohms} = \frac{\text{potential difference in volts}}{\text{current in amps}}$$

Resistance = _____ ohms

(c) Suggest one way in which Josh could improve his estimation of the resistance of the piece of wire.

_____

_____

(d) The wire that Josh investigated was made from nichrome. Nichrome is a mixture of the elements nickel, iron and chromium.

   (i) What is the name of a substance such as nichrome, made from a mixture of metals?

_____

1 mark

1 mark

1 mark

1 mark

(ii) Josh measures the resistance of other metals.

He obtains the following results:

Metal	Resistance in ohms
Aluminium	0.04
Constantan	0.83
Copper	0.03
Iron	0.16

Which metal would be best to use for wiring? Explain your answer.

_____

_____

_____

(e) Josh replaces a piece of wire with a solenoid with an iron core to make an electromagnet.

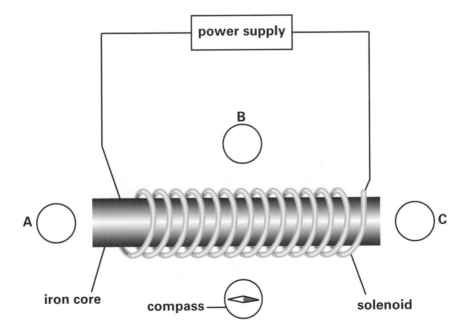

(i) What is a solenoid?

_____

(ii) Josh places a compass next to the solenoid, as shown.

He then moves the compass to positions A, B and C.

On the diagram above, show the direction each compass will point.

2. Three students have been growing some plants in the school greenhouse.

(a) They decide to measure how tall the plants have grown.

The diagram shows the position from which the students make their measurements.

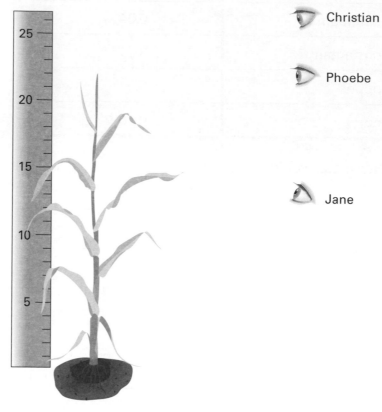

(i) The measurements they make of the plant's height are:

Student	Measurement made of height of plant in cm
Christian	21
Jane	24
Phoebe	22

Which student's measurement is likely to be closest to the true height of the plant? Explain your answer.

_____

_____

_____

_____

(ii) What do we call a measurement that is close, or identical, to its true value?

_____

(iii) The students go on to measure their other plants.

All the measurements made by two of the students have errors.

What is this type of error called?

_____

(b) Jane cuts a slice through one of her plant's leaves.

This is what she sees when she examines it with a microscope.

Give the name of:

Cell A _____

Structure B _____

(c) What is structure C called, and what is its function?

_____

_____

_____

(d) Write down two ways in which the leaf has adapted for photosynthesis.

_____

_____

_____

57

3. Ashraf is reacting some iron filings with dilute sulfuric acid.

(a) What gas is given off when iron and other metals react with an acid?

_____

(b) Ashraf measures the volume of gas produced. He tries out the experiment at different temperatures. He carries out each experiment for the same length of time.

Here are his results:

Temperature in °C	Volume of gas produced in cm³		
	Experiment 1	Experiment 2	Experiment 3
10	100	105	110
20	218	205	208
30	300	100	310
40	405	400	402
50	505	510	495

(i) Ashraf says that one of his results looks incorrect. Suggest which result this is.

_____

(ii) What do we call a result that is not consistent with the rest?

_____

(iii) At which temperature are the results the most repeatable?

_____

(c) Ashraf calculates the averages from his set of data.

Temperature in °C	Average volume of gas produced in cm³
10	105
20	210
30	305
40	402
50	503

(i) Plot a graph of these average volumes over temperature.

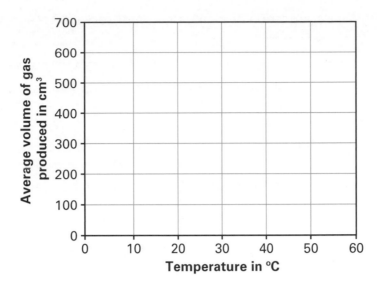

(ii) Draw a line of best fit of the data.

(iii) Describe how temperature affects the volume of gas produced.

_____

_____

_____

_____

(d) The mixture has many unreacted iron filings.

Ashraf would like to produce a pure solution of iron sulfate.

How should he produce this?

_____

_____

4. John is on holiday by the sea. He watches the waves as they move towards the seashore.

He thinks they slow down as they move towards the shore.

John reckons that the depth of water affects the speed of a wave.

1 mark

(a) What do we call a suggestion or idea based on observations?

_____

2 marks

(b) When John returns home, he tests this idea. He sets up a tank of water and produces a wave. The tank is 200cm long.

He times how long it takes for the wave to travel up the tank.

He tries the experiment with different depths of water. Here are his results. One has been done for you.

Depth of water in cm	Average time taken in seconds	Average speed in cm per second
1	6.5	
2	4.5	
3	3.8	
4	3.3	
5	3.2	63

Calculate the speed of the wave at the different depths and complete the table above.

1 mark

(c) To obtain his average results, John did the experiment three times.

Explain why scientists repeat experiments.

_____

1 mark

(d) What can you say about the results in relation to John's idea?

_____

_____

_____

5. Saima is investigating the solubility of different chemical compounds in water.

She investigates how much of each will dissolve at different temperatures.

(a) What term do we give to a chemical that:

    (i) dissolves in another chemical?

    _____

1 mark

    (ii) dissolves another chemical?

    _____

1 mark

(b) She draws a graph of her results.

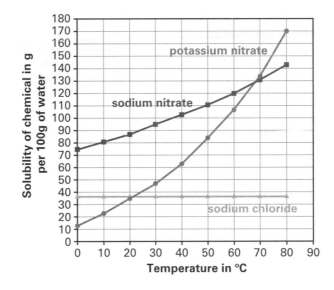

    (i) What is the maximum mass of sodium nitrate that will dissolve in 100g water at 50°C?

    _____

1 mark

    (ii) How does temperature affect the solubility of sodium chloride in water?

    _____

1 mark

    (iii) At what temperature does the solubility of potassium nitrate match the solubility of sodium chloride?

    _____

1 mark

(c) Potassium nitrate and sodium nitrate are both used as plant fertilisers.

In Saima's greenhouse, at 25°C, suggest which fertiliser would be better in terms of its solubility in water.

_____

1 mark

SUBTOTAL

6. Asif is studying a film of a game of cricket.

When a bowler bowls the ball, he times how long the cricket ball takes to travel the length of the wicket.

Bowler	Time to travel length of wicket in seconds	Bowling speed in metres per second
Graeme	0.74	27
Jimmy	0.50	
Mitchell	0.48	
Monty	0.85	
Ryan	0.51	
Stuart	0.49	

3 marks

(a) A cricket pitch is 20m long. Complete the table above by calculating the bowling speed of each of the bowlers.

2 marks

(b) Asif's own bowling speed is 27 metres per second.

Calculate Asif's bowling speed in kilometres per hour.

_____

7. In the science lab, Onnicha is analysing the air she breathes in, and the air she breathes out.

(a) These are her results:

Air	Oxygen concentration in %	Carbon dioxide concentration, in %
Air breathed in	20.90	0.04
Air breathed out	16.00	4.00

1 mark

(i) Compare the carbon dioxide concentration in the air breathed out with the air breathed in.

By how many times has it increased?

_____

(ii) What body process has led to this increase?

1 mark

_____

(iii) Write a word equation for this process.

2 marks

_____

_____   +   _____   →   _____   +   _____   (+ ENERGY)

(b) The carbon dioxide concentration in the air 50 years ago was 0.03%.

2 marks

Explain why it has increased.

_____

_____

_____

8. Steve has looked up car brake systems on the Internet. He finds a webpage that includes a diagram of how car brakes work.

The system shown includes a liquid called brake fluid. There are two pistons at each end. One has a larger area than the other.

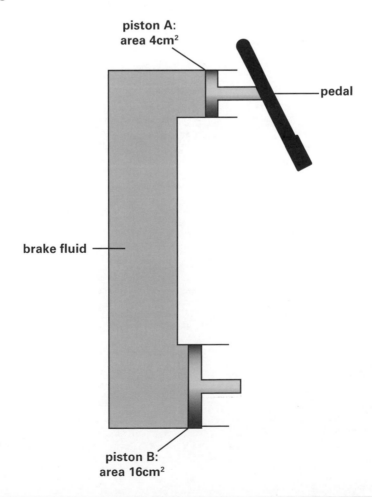

SUBTOTAL

1 mark

(a) Give the formula that shows the relationship between pressure, force and area.

____

2 marks

(b) (i) If a person applies a force of 240N on the brake pedal, calculate the pressure of piston A on the brake fluid.

____

____

1 mark

(ii) Assume that all this pressure is applied to piston B.
Calculate the force, in Newtons, on piston B.

____ N

2 marks

(c) Explain why car brakes are filled with liquid and not a gas.

____

____

____

9. The diagram below shows a pregnant woman's abdomen.

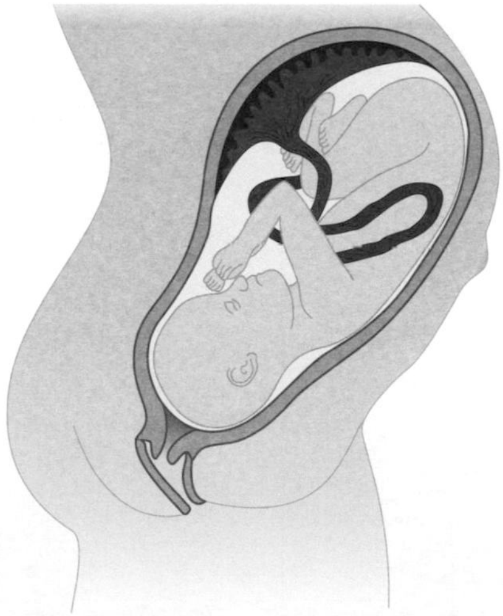

3 marks

(a) On the diagram, label the:

- amniotic fluid
- placenta
- umbilical cord
- uterus.

(b) How does the foetus receive its food and oxygen?

3 marks

_____

_____

_____

(c) Some pregnant teenage girls are admitted to hospital because of substance abuse.

The data below show percentages of girls admitted to a hospital in the USA in 1992 and 2007.

(i) Describe the changes in percentages of girls admitted for alcohol and cannabis abuse in 1992 and 2007.

2 marks

_____

_____

_____

_____

_____

(ii) Write down one effect on the foetus of a mother drinking alcohol.

1 mark

_____

_____

SUBTOTAL

10. In the past few years, scientists have produced certain chemicals as tiny particles called nanoparticles.

Scientists are now putting nanoparticles into sunscreens.

This person is wearing a normal sunscreen and a nanoparticle sunscreen.

before adding sunscreen

nanoparticle sunscreen

normal sunscreen

(a) Give **one** advantage of using nanoparticle sunscreen.

_____

_____

(b) Anna is testing how well different types of sunscreen protect against the Sun's rays. She covers pieces of plastic film with each of the sunscreens she is testing.

She places each piece of plastic on a sheet of Sun-sensitive paper.

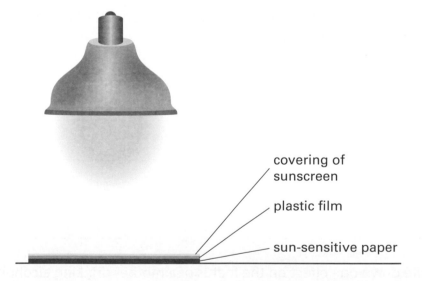

covering of sunscreen

plastic film

sun-sensitive paper

She turns on a sunlamp and records the time taken for the paper to change colour.

(i) Give **two** ways in which Anna can make this a fair test.

_____

_____

_____

(ii) How should Anna set up a control for the experiment?

_____

_____

(c) Some nanoparticles are made from zinc oxide.

Give **one** method of producing zinc oxide.

_____

_____

11. Anna, Laura and Mark are practising their archery. They are aiming for the bull's eye. They shoot some arrows into the target.

Anna's arrows   **Laura's arrows**   **Mark's arrows**

(a) Which statement best describes Anna's arrows? Tick the correct box.

Accurate and precise ☐   Precise but not accurate ☐

Accurate but not precise ☐   Neither precise nor accurate ☐

(b) Which statement best describes Laura's arrows? Tick the correct box.

Accurate and precise ☐   Precise but not accurate ☐

Accurate but not precise ☐   Neither precise nor accurate ☐

(c) Write a statement that best describes Mark's arrows.

_____

Set

C

KEY STAGE 3

Test 1

Science

Test Paper 1

Test Paper 1

First name   _____

Last name   _____

Date   _____

Instructions:

- The test is **75 minutes** long.

- Find a quiet place where you can sit down and complete the test paper undisturbed.

- You will need a pen, pencil, rubber and ruler. You may find a protractor and a calculator useful.

- The test starts with easier questions.

- Try to answer all of the questions.

- The number of marks available for each question is given in the margin.

- Show any rough working on this paper.

- Check your work carefully.

- Check how you have done using pages 101–112 of the Answers and Mark Scheme.

MAXIMUM MARK	75	ACTUAL MARK	

1. The local reservoir has turned bright green. Adrian, from the water company, is called in to investigate.

Adrian collects a water sample. He then looks at it with a microscope. This is what he sees.

(a) Adrian looks at the structure of the cells in the organism.

    (i) Which part of the cell is the cell membrane? Tick the correct box.

        A ☐   B ☐

        C ☐   D ☐

1 mark

    (ii) Which part of the cell is the nucleus? Tick the correct box.

        A ☐   B ☐

        C ☐   D ☐

1 mark

(b) Suggest **two** reasons why Adrian knows that the organism is a plant.

2 marks

_____

_____

_____

(c) Suggest an explanation for the reservoir having turned bright green.

2 marks

_____

_____

_____

SUBTOTAL

2. This diagram shows white light passing through a prism.

(a) (i) Add labels to the diagram to show the colours missing from the spectrum.

(ii) Which term shows what is happening to the light as it passes through the prism? Tick the correct box.

Absorption ☐   Dispersion ☐

Reflection ☐   Subtraction ☐

(b) In the days before Isaac Newton, people thought the colours came from the prism.

Describe an experiment to show that the colours of the spectrum come from the light, and not the prism.

_____

_____

_____

_____

_____

_____

3 marks

1 mark

2 marks

3. Rio runs his car on liquefied petroleum gas (LPG).

Many other people use LPG for their central heating or for portable appliances such as camping stoves.

LPG is supplied pressurised as a liquid, but burns as a gas.

(a) LPG contains the hydrocarbons propane and butane. What kind of chemical is LPG?

1 mark

Tick the correct box.

Element ☐   Compound ☐

Impure ☐   Mixture ☐

(b) The diagram below shows the properties of propane and butane.

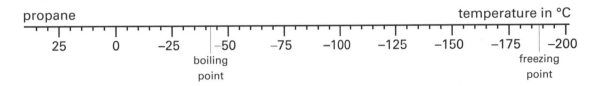

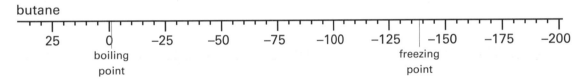

(i) At –25°C, in which state are propane and butane?

1 mark

Propane is a _____

Butane is a _____

(ii) At –145°C, in which state are propane and butane?

1 mark

Propane is a _____

Butane is a _____

(c) When propane (and butane) burns, it uses oxygen in the air to produce carbon dioxide and water. Write a word equation for the combustion of propane.

1 mark

_____

_____ + _____ → _____ + _____

SUBTOTAL

71

(d) Some gas suppliers change the proportions of butane and propane in the summer and winter. Suggest **two** reasons why they do this.

_____

_____

_____

4. The diagram below shows the structure of the human digestive system.

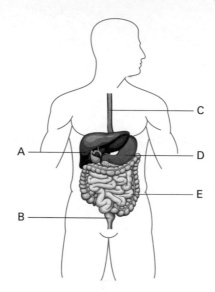

(a) (i) Which letter on the diagram labels the oesophagus? Tick the correct box.

A	B	C	D	E

(ii) Which letter on the diagram labels the liver? Tick the correct box.

A	B	C	D	E

(iii) Which letter on the diagram labels the small intestine? Tick the correct box.

A	B	C	D	E

(b) What is the name of the part of the digestive system:

    (i) in which the digestion of protein begins?

1 mark

_____

    (ii) where most of our food is absorbed into the blood?

1 mark

_____

(c) What **type** of chemical speeds up the digestion of our food? Tick the correct box.

1 mark

Acid   ☐   Enzyme   ☐

Saliva   ☐   Starch   ☐

5. Jenny is replacing her electric filament light bulbs with energy-saving bulbs.

She reads about the energy efficiency of the bulbs.

$$\text{energy efficiency in per cent} = \frac{\text{useful energy transferred}}{\text{energy supplied}} \times 100$$

(a) A diagram shows her the energy transfer in a filament light bulb.

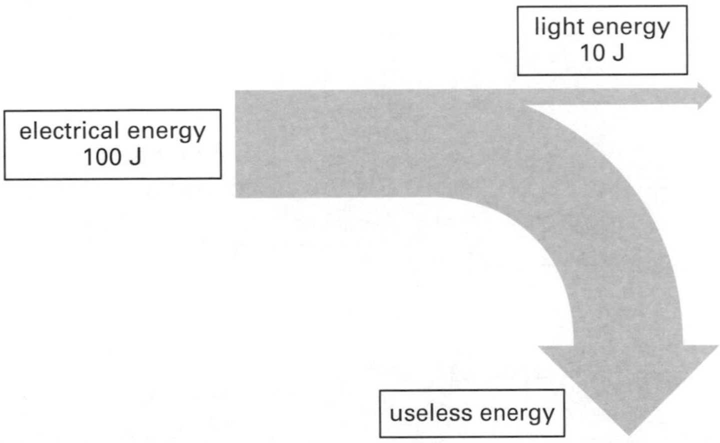

light energy
10 J

electrical energy
100 J

useless energy

    (i) How much energy is transferred to useless energy?

1 mark

_____ J

    (ii) In what form, or forms, is this useless energy?

1 mark

_____

    (iii) Calculate the energy efficiency of the light bulb.

1 mark

_____

SUBTOTAL

(b) An energy-saving bulb is supplied with 1500J of energy in one minute.

600J of energy is transferred to light.

Calculate the energy efficiency of the bulb.

_____

(c) Jenny is looking at the power used by some of her electrical appliances.

Complete the table:

Appliance	Power in W	Power in kW	Time appliance is used for in hours	Units of electricity used in kWh
Hair dryer		2.0	1	
Kettle	3000		2	
Light bulb	100		20	
Toaster		1.3	1	

6. The diagram shows the structure of the human arm.

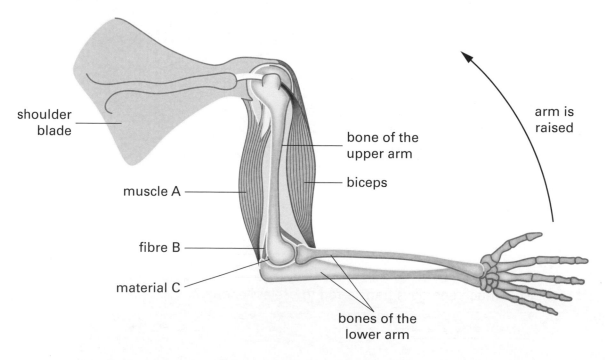

(a) (i) Give the name of muscle A.

_____

(ii) What happens to the biceps when it is used to raise the arm?

_____

(iii) Explain why muscles of the arm work in pairs.

2 marks

_____

_____

_____

(b) Describe the function of:

2 marks

Fibre B: _____

Material C: _____

7. Sam is using some nail polish remover. It contains pentyl ethanoate and propanone.

(a) The formula of pentyl ethanoate is:

(i) Which elements are present in a molecule of pentyl ethanaote?

1 mark

_____

(ii) How many atoms are present in a molecule of pentyl ethanoate?

1 mark

_____

(b) Pentyl ethanoate is irritant. Propanone is highly flammable and irritant.

(i) Which symbol, or symbols, would you find on a bottle of pentyl ethanoate?

1 mark

Tick the correct box(es).

SUBTOTAL

(ii) Which symbol, or symbols, would you find on a bottle of propanone?

Tick the correct box(es).

☐   ☐   ☐   ☐   ☐

1 mark

(iii) Give one safety precaution Sam should use when using her nail polish remover.

_____

_____

_____

1 mark

8.   Zoe is a research scientist in the car industry. She is investigating the motion of a car.

(a) Zoe is test driving the car.

    (i) Complete the following sentence.

       Force from the car's _____ is causing the forward movement of the car.

1 mark

    (ii) Name **two** forces that are resisting the movement of the car.

_____

_____

2 marks

(iii) Zoe drives the car at a steady speed around the track.

What can you say about the forces on the car?

_____

_____

(b) Zoe is testing the car on different road surfaces on the track.

She measures the car's fuel use on the different road surfaces.

Type of road surface	Fuel use in cm³ fuel used per km
Tarmac® with small chippings	67
Tarmac® with large chippings	69
Concrete with small chippings	70
Concrete with large chippings	72

Draw a bar chart that shows the car's fuel use on different road surfaces.

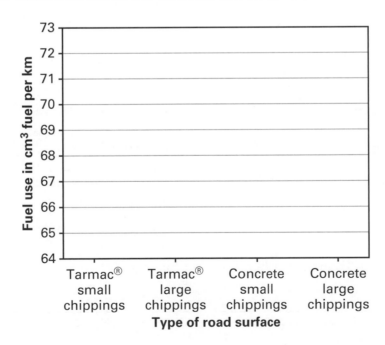

9. The picture below shows Aiysha's kitchen.

(a) Several types of material are shown in the picture.

(i) What type of material are wall tiles made from? Tick the correct box.

Ceramic ☐   Composite ☐

Metal ☐   Polymer ☐

(ii) What type of material is the plastic storage jar made from? Tick the correct box.

Ceramic ☐   Composite ☐

Metal ☐   Polymer ☐

(b) Stainless steel is an alloy. What is meant by an alloy?

_____

_____

(c) Describe how granite is formed.

_____

_____

10. Vicky is investigating electromagnets.

She winds a coil of insulated copper wire around a rod of pure iron. She then connects the wire to the power supply.

She counts how many paperclips the electromagnet will pick up with different numbers of turns in the coil.

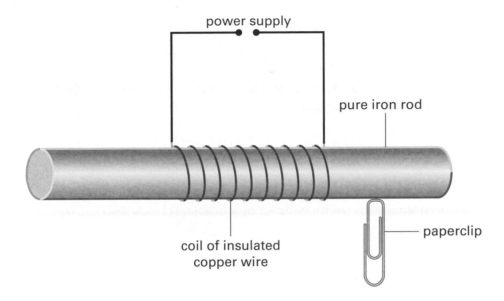

(a) Her results are shown below:

Number of turns in the coil	Number of paperclips picked up
10	1
20	2
30	3
40	4
50	5
60	6

(i) Describe the relationship between the number of turns in the coil and the strength of the electromagnet.

_____

_____

_____

1 mark

(ii) Give one other way in which she could increase the strength of the electromagnet.

_____

1 mark

SUBTOTAL

(b) Vicky turns the power supply off.

2 marks

(i) What would happen to the paperclip(s)? Explain your answer.

_____

_____

_____

1 mark

(ii) Vicky repeats this with a steel rod instead of the iron rod. Explain why the paperclips remain attached.

_____

11. When a pollen grain lands on the stigma of a flower, it begins to grow.

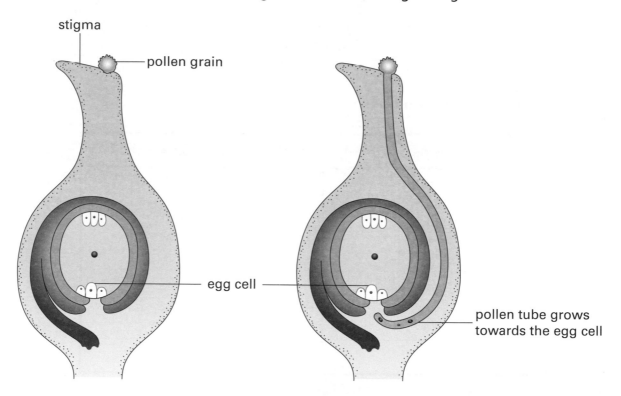

Jodie collects some pollen grains from a spider flower.

She places them in sugar solution on a microscope slide. She watches them as they grow.

She investigates the effects of several different concentrations of sugar solution.

Here are her results:

Concentration of sugar solution in %	Pollen grains that have grown after one hour in %
0	0
5.0	43
7.5	67
10.0	88
12.5	83
15.0	72

(a) Plot a graph of Jodie's results. Draw a line of best fit through the points.

3 marks

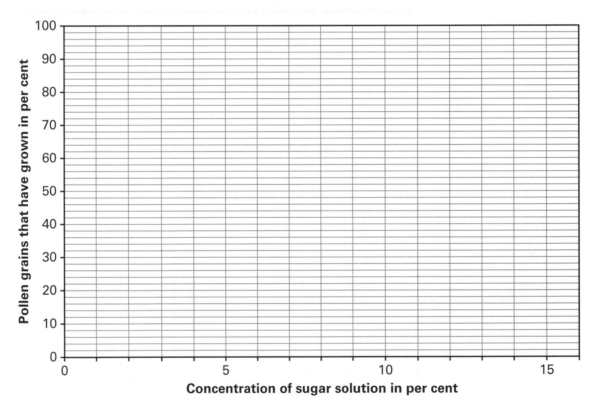

(b) In Jodie's investigation:

(i) Give the dependent variable.

1 mark

_____

(ii) Give one variable that she kept constant.

1 mark

_____

SUBTOTAL

(c) Describe the pattern of her results.

_____

_____

_____

_____

(d) Suggest one reason why the pollen grains needed sugar to grow.

_____

_____

12.   Erica is going to recycle her mobile phone.

She looks at some information on the Internet on the metals used in mobile phones.

Element	Total mass in the world's mobile phones in 2010, in tonnes	Concentration in Earth's crust, in part per million (ppm)*
Aluminium	4200	82000
Copper	21000	50
Gold	140	0.0011
Iron	4200	41000
Palladium	30	0.0006
Platinum	1	0.001
Silver	700	0.07
Tantalum	28	2

*Number of parts of the element per million parts of Earth's crust

(a) Which metal is most used in the world's mobile phones?

_____

(b) Using information from the table:

    (i) Which is the **most** abundant element in the Earth's crust? Tick the correct box.

       Aluminium ☐   Gold ☐

       Iron ☐   Palladium ☐

    (ii) Which is the **least** abundant element in the Earth's crust? Tick the correct box.

       Aluminium ☐   Gold ☐

       Iron ☐   Palladium ☐

(c) Erica says that it's more important to recycle metals such as gold from mobile phones than it is the aluminium.

    (i) Explain why.

_____

_____

_____

    (ii) Erica reads that to recycle aluminium, the cost is only 5% of that needed to extract it from its ore.

       Give **two** other reasons why it's often better to recycle metals than to mine and extract more.

_____

_____

_____

Set **C**

KEY STAGE 3

Test 2

Science

Test Paper 2

Test Paper 2

First name _____

Last name _____

Date _____

Instructions:

- The test is **75 minutes** long.

- Find a quiet place where you can sit down and complete the test paper undisturbed.

- You will need a pen, pencil, rubber and ruler. You may find a protractor and a calculator useful.

- The test starts with easier questions.

- Try to answer all of the questions.

- The number of marks available for each question is given in the margin.

- Show any rough working on this paper.

- Check your work carefully.

- Check how you have done using pages 101–112 of the Answers and Mark Scheme.

MAXIMUM MARK	75		ACTUAL MARK	

1. Sarah's teacher sets up a large glass tube.

In one end she puts a piece of cotton wool soaked in ammonia solution.

In the other end, she puts a piece of cotton wool soaked in concentrated hydrochloric acid.

cotton wool soaked in ammonia   cloud of chemical particles   cotton wool soaked in hydrochloric
solution gives off ammonia gas   acid gives off hydrogen chloride gas

(a) The cotton wool soaked in ammonia solution gives off ammonia gas.

The cotton wool soaked in concentrated hydrochloric acid gives off hydrogen chloride gas.

(i) Explain why particles of ammonia gas and hydrogen chloride gas move along the glass tube.

2 marks

_____

_____

_____

(ii) What is this process called?

1 mark

_____

(iii) What are the particles of ammonia and hydrogen chloride called?

1 mark

_____

SUBTOTAL

(b) Within a few seconds of putting the pieces of cotton wool in the glass tube, a white cloud forms part way along the tube.

(i) What has happened to produce the cloud of chemical particles?

_____

_____

(ii) Suggest why the cloud forms closer to the hydrogen chloride end of the tube and not in the middle.

_____

_____

2. Tony is laying some slabs in his garden. They are made from limestone.

(a) Each slab is 0.3m × 0.3m and weighs 72N.

Calculate the pressure that would be exerted on the ground by one slab.

_____ N/m²

(b) Limestone is a natural material.

(i) What type of rock is limestone?

1 mark

_____

(ii) How is this type of rock formed?

2 marks

_____

_____

_____

_____

(c) Marble is a rock that is formed from limestone.

(i) What type of rock is marble?

1 mark

_____

(ii) How is marble formed from limestone?

1 mark

_____

_____

87

3. Coral reefs are important marine ecosystems.

(a) What is meant by an ecosystem?

_____

_____

1 mark

(b) Carbon dioxide reacts with water to form carbonic acid.

Write a word equation for this reaction.

_____

_____ + _____ → _____

2 marks

(c) Emissions of carbon dioxide and other gases are causing the sea to become more acidic.

Acidic seawater destroys coral.

(i) Give **one** source of emissions of carbon dioxide.

_____

1 mark

(ii) The graph opposite shows how the pH of seawater has changed since the first measurements were made in 1850.

Values of pH shown on the graph have been compared with those of 1850.

2 marks

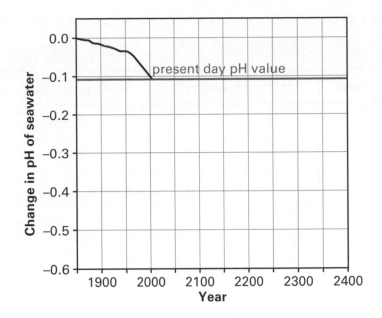

Describe the change in pH from 1850 to the present day.

_____

_____

_____

(iii) What is the overall change in pH during this time?

1 mark

_____

(iv) Extending the line, use the graph to predict the year when the pH will have decreased by a value of 0.5 since 1850.

2 marks

_____

(v) What did you assume when you made this prediction?

1 mark

_____

_____

SUBTOTAL

4. The table below gives the properties of Group 7 elements of the Periodic Table. They are in group order.

Group 7 element	Melting point in °C	Boiling point in °C
Fluorine	−219	−188
Chlorine	−101	−34
Bromine	−7	59
Iodine	114	184
Astatine	302	380

(a) (i) Which elements are solid at room temperature?

_____

(ii) Which element is liquid over the greatest temperature range?

_____

(b) The properties of elements change as you move down (or up) a group.

(i) Using information from the table, describe how **one** property changes as you move down the Group 7 elements.

_____

_____

_____

(ii) Using other information, not in the table, describe how **one** property changes among the Group 7 elements.

_____

_____

_____

_____

1 mark

1 mark

1 mark

1 mark

(iii) Predict what would happen if you added chlorine (in solution) to a solution of potassium bromide.

1 mark

_____

_____

5. Polar bears first appeared around 200 000 years ago.

It is thought that they evolved from brown bears stranded in the Arctic.

Polar bear   Brown bear

(a) Brown bears show a lot of variation in their coat colour.

(i) What is this type of variation called?

1 mark

_____

(ii) Suggest why brown bears with pale coats would be more likely to survive if stranded in the Arctic.

2 marks

_____

_____

(b) Suggest how polar bears may have developed as a new species.

2 marks

_____

_____

SUBTOTAL

6. The Moon is the Earth's satellite and orbits our planet.

We see the part of the Moon that is lit up by the Sun.

1 mark
(a) (i) How long does It take for the Moon to orbit the Earth?

_____

1 mark
(ii) How is the Moon kept in orbit around the Earth?

_____

(b) This diagram shows the Sun, Earth and Moon at different times as the Moon orbits the Earth.

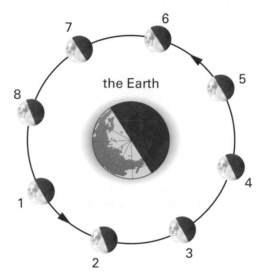

the Sun

the Moon orbiting the Earth

1 mark
(i) In which number, or phase, on the diagram is the Moon at its most visible to us?

_____

2 marks
(ii) The diagram below shows some images of the Moon observed from the Earth.

Referring to the first diagram also, give the letter which shows what we would see from Earth:

At phase 1: _____

At phase 7: _____

7. On a sunny day, Josh notices that the waterweed in his aquarium produces many bubbles of a gas. On a cloudy day, there are very few bubbles of the gas.

(a) Suggest a hypothesis to explain these observations.

2 marks

_____

_____

_____

(b) Josh decides to investigate the gas.

He puts several pieces of pondweed in a beaker of water, and places a boiling tube that he has filled with water over the funnel.

He stands the apparatus on the window sill for several days.

gas collected

water containing sodium hydrogencarbonate

filter funnel

water weed

(i) Josh adds sodium hydrogencarbonate, $NaHCO_3$, to the water. Suggest why he does this.

1 mark

_____

_____

(ii) What process leads to the production of the gas collected?

1 mark

_____

SUBTOTAL

(c) Write a word and symbol equation for this process.

(i) Word equation:

$$\underline{\hspace{3cm}} + \underline{\hspace{3cm}} \xrightarrow[]{\text{light energy}} \underline{\hspace{3cm}} + \underline{\hspace{3cm}}$$

(ii) Symbol equation:

$$\underline{\hspace{3cm}} + \underline{\hspace{3cm}} \rightarrow \underline{\hspace{3cm}} + \underline{\hspace{3cm}}$$

8. Ben wants to test the energy content of some different types of fuels in the lab. He then:

- pours some of the first fuel into the spirit burner

- sets the fuel alight and lets it burn

- then measures the temperature rise of the water

- repeats the same experiment to test each fuel.

(a) Give **two** variables that Ben must keep constant.

thermometer

boiling tube (held in clamp)

water

spirit burner

fuel

_____

_____

_____

(b) Before Ben begins the experiment, he needs to fill in a Risk Assessment form.

(i) Complete the Risk Assessment for the fuel by filling in the gaps.

Chemical	Hazard	Safety precaution
Fuel	Flammable	
	Toxic vapour	

(ii) Name **one** other hazard in this experiment.

_____

2 marks

2 marks

2 marks

1 mark

1 mark

(c) Here are Ben's results:

Type of fuel	Temperature rise in °C			Average rise in temperature in °C
	Test 1	Test 2	Test 3	
Ethanol	16.5	17.0	16.8	
Propanol	19.0	18.8	19.3	
Butanol	20.5	21.2	21.0	

(i) Calculate the average rise in temperature for each fuel.
Fill in the end column of the table.

(ii) Which fuel releases most energy when burned?

_____

(d) What is the name given to chemical reactions that give out heat?

_____

9. Helena is a marine biologist. She is studying the hearing range of some marine animals.

(a) The graph below shows the hearing range of two animals.

It shows the limits of their hearing across a range of frequencies.

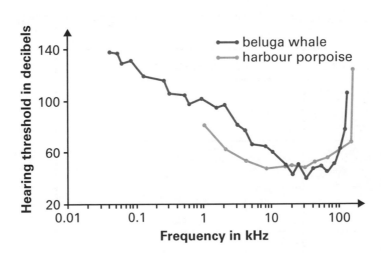

(i) What is the quietest sound that can be heard by the beluga whale?

_____

(II) What is the lowest pitch sound that can be heard by the harbour porpoise?

_____

(b) The hearing range of most humans is 20–20 000Hz.

(i) Mark this range on the graph on p 95.

(ii) Compare the hearing range of humans with that of the beluga whale and harbour porpoise.

_____

_____

_____

_____

_____

_____

_____

10. Sulfuric acid is made on a large scale. It is one of our most important chemicals.

(a) The table below shows some products made using sulfuric acid.

Product made	% of sulfuric acid used
Detergent	19
Dyes	7
Fertilisers	4
Fibres	3
Paint	20
Plastics	4

Draw a bar chart to show these uses.

Product

(b) Sulfuric acid is made in two stages:

Stage 1: sulfur dioxide is reacted with oxygen to form sulfur trioxide.

Stage 2: sulfur trioxide is reacted with water to form sulfuric acid.

(i) What type of reaction is stage 1?

_____

(ii) The symbol equation for the stage 1 reaction is:

2 marks

$$2SO_2 \quad + \quad O_2 \quad \rightarrow \quad 2SO_3$$

Tick two correct boxes:

Statement about the reaction	Tick (✔)
The mass of the product is two-thirds that of the reactants	
The mass of the product is the same as the mass of the reactants	
Two molecules of sulfur trioxide are formed from one atom of oxygen	
One molecule of sulfur trioxide is formed from one molecule of sulfur dioxide	

(iii) The formula of sulfuric acid is H_2SO_4.

2 marks

Write a symbol equation for **stage 2** of the reaction.

_____   +   _____   →   _____

11. The graph below provides some data on men who have smoked in the UK, from 1948 to 2010.

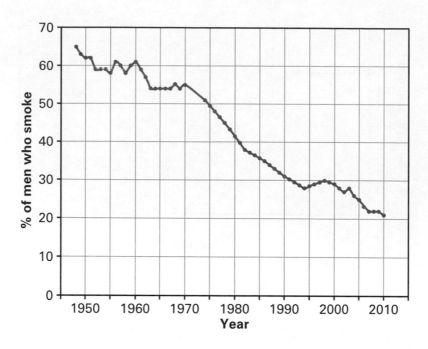

(a) Describe the trends in smoking in men from 1948 and 2010.

2 marks

_____

_____

_____

(b) The following data show the number of cases of lung cancer in men between 1975 and 2010.

Year	Number of lung cancer cases in the UK*
1975	113
1980	113
1985	110
1990	94
1995	82
2000	71
2010	58

* numbers per 100 000 of the population

(i) Describe how the trend in lung cancer cases suggests a link with the trend in smoking.

_____

_____

_____

(ii) Why must scientists be careful when interpreting this kind of data?

_____

_____

12. Jason is a forensic scientist.

He is analysing a chemical sample from a crime scene. He thinks that it could be the drug cocaine.

On a piece of filter paper, he places:

- a chemical sample from the crime scene
- a sample of pure cocaine on a piece of filter paper.

He then places the filter paper in a solvent for one hour.

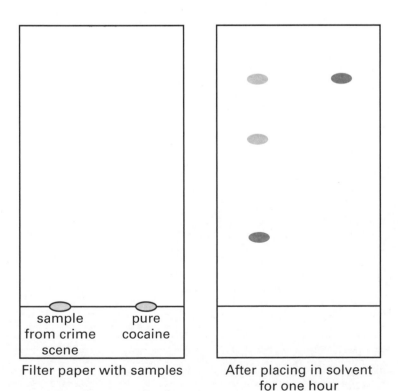

Filter paper with samples   After placing in solvent for one hour

(a) Give a definition of a solvent.

_____

_____

(b) State the name of the technique that Jason is using to analyse the sample.

_____

(c) Explain why Jason included a sample of pure cocaine on the filter paper.

_____

_____

_____

(d) Write down two things that Jason can conclude about the sample from the crime scene.

1 _____

_____

2 _____

_____

Answers and Mark Scheme

Set A, Test Paper 1

1. **(a)** **(i)** Electron *(1 mark)*
    (ii) Neutron *(1 mark)*
    (iii) Proton *(1 mark)*
  (b) Electron 1/1800
    Neutron 1 *(1 mark)*
    Helpful hint: *Note that in this case, two answers are required for the mark.*

2. **(a)** **(i)** **Two** answers from: blue eyes; dimple; natural hair colour; straight hair *(1 mark)*
    (ii) **Two** answers from: decayed tooth; pierced ear; pink hair; scar *(1 mark)*
    Helpful hint: *It would also be acceptable here to include her skin colour or hairstyle.*
  (b) Hair colour; skin colour *(1 mark)*
  Helpful hint: *It is best to describe the characteristics, rather than how they appear in Jade.*
  (c) **(i)**

    Scale and labelled axes *(1 mark)*
    Bars drawn correctly *(1 mark)*
    (ii) Blood groups: discontinuous variation *(1 mark)*
    Height: continuous variation *(1 mark)*

3.

Torch	**chemical/ electrical**	light	**heat**	*(1 mark)*
Bunsen burner	**chemical**	**heat**	light	*(1 mark)*
Stretched spring	elastic potential	**kinetic**	**heat/ sound**	*(1 mark)*
Toy car	**(gravitational) potential**	kinetic	**heat/ sound**	*(1 mark)*

Helpful hint: *Examiners often award one mark per row in this type of question.*
You will not have to include all the wasted forms of energy to gain the mark.

4. **(a)** **(i)** C *(1 mark)*
    (ii) B *(1 mark)*

(b) **(i)** *(1 mark)*
    (ii) *(1 mark)*

ovary
vagina

  (c) The testes and the ovaries *(1 mark)*
  (d) The Fallopian tubes or oviduct *(1 mark)*
  Helpful hint: *Because the embryo and foetus develop in the uterus, students sometimes think that fertilisation occurs here also; it doesn't.*

5. **(a)** **(i)** 5 (two iron and three oxygen) *(1 mark)*
    (ii) It is becoming oxidised *(1 mark)*
  (b) Aluminium is more reactive/higher in the reactivity series than carbon; *(1 mark)* so will remain combined with the oxygen (if heated with carbon) *(1 mark)*
  (c) Gold and silver are very unreactive metals (so will not react with chemicals such as oxygen, water, acids, etc., in the environment); *(1 mark)* sodium is a very reactive metal (so will react quickly with chemicals such as oxygen, water, acids, etc., in the environment) *(1 mark)*

6. **(a)** A Blue
    B Green
    C Red *(1 mark)*
  (b) **(i)** Red *(1 mark)*
    (ii) Black *(1 mark)*
    Helpful hint: *A blue object is blue because it reflects blue light. It cannot reflect the red light, so reflects no light at all. It appears black.*
    (iii) Red *(1 mark)*
    Helpful hint: *A yellow object is yellow because it reflects red and green light. If under red light, it will reflect red light only, so will appear red.*

7. **(a)** A, B, D, F and G *(2 marks)*
    (2 marks if all correct; 1 mark for two, three or four answers; no marks for one correct)
  (b) A, B, D, F and G *(2 marks)*
    (2 marks if all correct; 1 mark for two, three or four answers; no marks for one correct)
  Helpful hint: *The same as (a) (i). These are combustion (A and B), and respiration (D, F and G).*
  (c) C *(1 mark)*
  (d) Combustion (or 'burning') *(1 mark)*

8. **(a)** **One** from: no oxygen in space/little or no air pressure/body fluids/blood would boil/tissues expand/harmful radiation in space/low temperatures *(1 mark)*

space suit provides oxygen/pressurised/
protective environment *(1 mark)*
- **(b)** The astronaut will not move in any direction/
will be stationary; *(1 mark)*
as upward and downward forces are equal, and
left and right forces are equal *(1 mark)*
*Helpful hint: It's not quite enough to say that
forces are balanced, as the sideways forces are not
the same values as the up and down forces.*
*Helpful hint: You are not required to give values of
the forces (these are only an illustration), and should
just give the direction of the forces.*
- **(c)** 160N *(1 mark)*
*Helpful hint: Remember that weight is expressed
in Newtons.*
For this calculation, you need the formula:
weight = mass × gravitational field strength.

9. (a) DNA *(1 mark)*
- **(b)** (Francis) Crick; (James) Watson; (Rosalind)
Franklin; (Maurice) Wilkins
*(2 marks if all correct; 1 mark for two or three; no
marks for one correct)* *(2 marks)*
- **(c)** Nucleus *(1 mark)*
- **(d)** Our DNA is unique to us (unless we are an
identical twin); *(1 mark)*
so DNA left at a crime scene will be evidence
that a person was there *(1 mark)*

10. (a) (Dmitri) Mendeleev *(1 mark)*
- **(b)** Elements with similar properties grouped
together; *(1 mark)*
arranged in order of atomic mass *(1 mark)*
- **(c)** Non-metals *(1 mark)*
- **(d)** Group *(1 mark)*

11. (a) The temperature stays the same; *(1 mark)*
as energy absorbed; *(1 mark)*
as the hexadecanoic acid melts *(1 mark)*
- **(b)** Clear horizontal line (as hexadecanoic
acid changes state); *(1 mark)*
if pure *(1 mark)*
(or the opposite: there is no clear horizontal line
(1 mark) if impure *(1 mark)*)

12. (a) Sound needs a medium/particles through
which to travel/it cannot travel in a vacuum
(or with no air present) *(1 mark)*
*Helpful hint: You need to be careful with how you
express your answer. It's not correct to say that
sound needs air to travel through; it can travel
through other gases, along with liquids and solids.*

(b) (i)

Scale and labelled axes *(1 mark)*
Five or six bars drawn correctly *(2 marks)*
Three or four bars drawn correctly *(1 mark)*
(Maximum 3 marks)
- **(ii)** Sound travels most quickly in solids, followed
by liquids, slowest in gases; *(1 mark)*
vibrations are transferred more quickly if
particles are closer together *(1 mark)*
- **(c)** Sound transmitted towards sea bed; *(1 mark)*
reflected back to ship; *(1 mark)*
distance is calculated from speed of sound and
time taken *(1 mark)*

13. (a) (i) Ammeter *(1 mark)*
- **(ii)** 0.5A *(1 mark)*
*Helpful hint: Remember that current is not used
up in a circuit, so will be the same at all places in
the circuit.*
- **(b) (i)** Parallel *(1 mark)*
- **(ii)** M_4 0.2A
M_5 0.6A *(1 mark)*
*Helpful hint: Remember that current will be split
along the three routes in this parallel circuit. In
this case, it's divided equally, but may not be if the
components along the three routes are different.*

14. (a) (i) The concentration is not sufficient to kill the
microscopic plankton and fish; *(1 mark)*
but the concentration builds up as it is passed
along the food chain /insecticide accumulates
because birds eat many fish; *(1 mark)*
to poisonous levels *(1 mark)*
- **(ii)** Humans will eat the fish *(1 mark)*

Set A, Test Paper 2

1. (a) There may be some variation (in content) between different bars *(1 mark)*
    (b) Calculation showing amount per bar ÷ amount per 100g × 100 *(1 mark)*
        40(g) *(1 mark)*
        Helpful hint: *You can do the calculation using values from any of the food types. They will all give the same answer.*
    (c) Protein: growth (and development and repair)
        Carbohydrate: energy
        Fat (lipids): energy/development *(1 mark each)*
    (d) Minerals *(1 mark)*
        Vitamins *(1 mark)*
    (e) 5 *(1 mark)*
        (20 ÷ 4)

2. (a) (i) Coal *(1 mark)*
        (ii) Solar and wind *(1 mark)*
            Helpful hint: *Note that both answers are required for the mark.*
        (iii) Nuclear, solar and wind *(1 mark)*
            Helpful hint: *Note that all three answers are required for the mark.*
        (iv) Solar and wind *(1 mark)*
    (b) (gravitational) potential energy → *(1 mark)*
        kinetic → *(1 mark)*
        electrical *(1 mark)*
        Helpful hint: *The question asks for energy transfers, so these should be in the correct order.*

3. (a) 20°C at 1000 million N/m² *(1 mark)*
    (b) −40°C *(1 mark)*
    (c) It increases *(1 mark)*
    (d) (i) B *(1 mark)*
        (ii) C *(1 mark)*
        (iii) E *(1 mark)*
    (e) (i) The point at which carbon dioxide exists as a solid, liquid and gas *(1 mark)*
        (ii) −57°C *(1 mark)*

4. (a) **Two** answers from the following:
        Tissue lining the inside of the mouth is made up of flattened cells/the lining tissue from the trachea has tall cells; *(1 mark)*
        tissue lining the inside of the mouth is made up of one type of cell/the lining tissue from the trachea is made up of more than one type; *(1 mark)*
        the lining tissue from the trachea has cilia; *(1 mark)*
        the tissue lining in the trachea produces mucus *(1 mark)*
        Helpful hint: *Make it clear which tissue, or the cells making up the tissue, have the features that you are referring to.*

    (b) Magnification = $\dfrac{\text{size on diagram}}{\text{actual size}}$ *(1 mark)*
        = $\dfrac{30}{0.06}$
        = × 500 *(1 mark)*

5. (a) Anticlockwise *(1 mark)*
    (b) Turning moment = weight × distance from the pivot
        350 × 2 *(1 mark)*
        700(Nm) *(1 mark)*
        Helpful hint: *When doing a calculation in a test, write out the formula and show your working out (in this question, you will be awarded one mark for this).*
    (c) Turning moment = weight × distance from the pivot
        700 = weight × 1.75
        ∴ weight = 700 ÷ 1.75 = 400 *(1 mark)*
        400(N) *(1 mark)*
        Helpful hint: *For the see-saw to be balanced, the anticlockwise and clockwise turning forces must be the same. Again, there is one mark for showing your calculation.*
    (d) 730 *(1 mark)*
        700 + (20 × 1.5) = 730Nm *(1 mark)*
        Helpful hint: *You add the moments to find the total anticlockwise turning moment.*

6. (a) (i)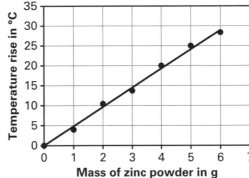
            Four or five points plotted correctly *(2 marks)*
            Two or three points plotted correctly *(1 mark)*
            (Maximum two marks)
        (ii) Line of best fit *(1 mark)*
            Helpful hint: *If the line of best fit does not pass through the points, you should make sure that there is an equal number of points above and below the line.*
    (b) Jess's experiment is more reliable *(0 marks)*
        Two of the following points:
        The polystyrene is a better insulator than glass; *(1 mark)*
        so less heat is lost (from the container); *(1 mark)*
        use of a lid reduces heat loss (from the container); *(1 mark)*
        and Jess gets a more accurate measurement of the temperature rise *(1 mark; maximum 2 marks)*

Helpful hint: *Note that you do not get a mark for saying who will get more reliable results. The marks are for your explanation.*

(c) copper sulfate + zinc *(1 mark)*

copper + zinc sulfate *(1 mark)*

Helpful hint: *In a chemical equation, examiners usually award 1 mark for getting the reactants correct, and one mark for the products.*

(d) Magnesium is higher in the reactivity series/ more reactive than zinc *(1 mark)*

7. (a) Yeast *(1 mark)*

(b) (i) $C_6H_{12}O_6 + 6O_2 \rightarrow$ *(1 mark)*

$6CO_2 + 6H_2O$ *(1 mark)*

(energy)

(ii) Glucose *(1 mark)*

carbon dioxide + alcohol *(1 mark)*

(energy)

(c) 1.05(g per cm³) *(1 mark)*

8. (a) (i) C *(1 mark)*

(ii) The Earth is tilted (at 23.5°) on its axis; *(1 mark)*

When it is winter in the Northern Hemisphere, this hemisphere is tilted away from the Sun, so the Southern Hemisphere is tilted towards the Sun, and it is summer. *(1 mark)*

Helpful hint *A very common misconception is that the Earth is closer to the Sun in the summer. This is wrong!*

(b) (i) One year/ 365 days/ 365.24 days *(1 mark)*

(ii) 24 hours *(1 mark)*

9. (a) acid + alkali *(1 mark)*

salt + water *(1 mark)*

(b) (i) The pH will increase; *(1 mark)*

until it becomes neutral/pH 7 *(1 mark)*

(ii) (Use an) indicator *(1 mark)*

(c) $KOH + HNO_3$ *(1 mark)*

$KNO_3 + H_2O$ *(1 mark)*

(d) Evaporate off most of the water; *(1 mark)*

and allow the fertiliser to crystallise *(1 mark)*

Helpful hint: *You would still get the two marks if you said evaporate off the water* (1 mark) *until the fertiliser is dry (1 mark), though you should not heat dry potassium nitrate in the lab.*

10. (a) 15 seconds *(1 mark)*

(b) (i) Becomes quicker *(1 mark)*

Becomes deeper/volume increases *(1 mark)*

(ii) She is taking in more oxygen by increasing breathing rate/lung volume; *(1 mark)*

oxygen is needed by muscles (for respiration) *(1 mark)*

11. (a) 200 seconds/ 3 minutes 20 seconds *(1 mark)*

(b) 7.5m/s *(1 mark)*

Helpful hint: *Use the formula speed = distance ÷ time, substituting the values into the formula: speed = 1500 ÷ 200.*

(c) After 80 seconds/ 1 minute 20 seconds/between 80 and 110 seconds into the journey *(1 mark)*

(d) (i) Between 110 seconds and 200 seconds (into the journey) *(1 mark)*

Helpful hint: *You can quickly identify this period by looking for the steepest gradient on the graph.*

(ii) 12.2m/s

Distance travelled during this time = (1500 − 400) m = 1100m *(1 mark)*

Time taken for this journey = (200 − 110) s = 90s *(1 mark)*

Speed = distance ÷ time = 1100 ÷ 90 = 12.2m/s *(1 mark)*

Set B, Test Paper 1

1. (a) (i) Displacement *(1 mark)*
Helpful hint: *It's also an oxidation, but you would not be expected to know this, at Key Stage 3 level.*
(ii) silver nitrate + copper *(1 mark)*
(→)
silver + copper nitrate *(1 mark)*
Helpful hint: *In a chemical equation, examiners usually award 1 mark for getting the reactants correct, and one mark for the products.*

(b) (i) Iron
Tin
Lead
Copper
(2 marks if all four metals in correct order; 1 mark if three or two metals in correct order)
(2 marks)
(ii) Below copper *(1 mark)*
Because copper displaces silver (from silver nitrate) *(1 mark)*

2. (a) (i) slows down *(1 mark)*
refraction *(1 mark)*
speeds up *(1 mark)*
(ii) Two answers from:
cameras; spectacles; telescope, binoculars, projector, lighthouse, magnifying glass/ hand lens; the eye *(1 mark)*
Helpful hint: *Note that in this question, you need to have two correct examples to get the mark.*

(b) Angle *i* equals angle *r*
A mirror image is reversed, left to right *(1 mark)*
Helpful hint: *Note that again, in this question, you need to have both statements correct to get the mark.*

3. (a) cell wall *(1 mark)*
chloroplast *(1 mark)*
nucleus *(1 mark)*
(b) (i) A and E *(1 mark)*
(ii) B *(1 mark)*
(iii) C *(1 mark)*
(iv) Mitochondria *(1 mark)*
(v) Diffusion *(1 mark)*

4. (a) (i) Newton *(1 mark)*
(ii) Stretch *(1 mark)*

(b) (i)

2 marks if all six points are plotted correctly
1 mark if three, four or five points are plotted correctly *(2 marks)*
(ii) Good line of best fit (see above) *(1 mark)*
Helpful hint: *In this case, the line of best fit should go through all the points.*
(iii) As the force is increased, there is an increase in the length of the spring; *(1 mark)*
this increase in length is directly proportional to the force/as the force is doubled, the increase in length of the spring doubles *(1 mark)*
Helpful hint: *You should have found the first mark straightforward to obtain, but the second is more difficult. Look out for these types of relationships in your data.*

5. (a) (i) C *(1 mark)*
(ii) A *(1 mark)*
(b) (i) The balloons *(1 mark)*
(ii) The diaphragm *(1 mark)*
(iii) Three of the following points:
Movement of the (rib cage) and diaphragm;
cause an increase in volume of the chest/ thorax;
lowering pressure (in the chest);
so air is breathed in *(3 marks)*

6. (a) Liquid   Solid   Gas *(1 mark)*
(b) Particles/molecules of the liquid gain in energy; *(1 mark)*
some have enough energy to turn into a vapour/ break through the surface of the liquid/to break forces between particles/so that they can move further apart *(1 mark)*
(c) In a gas, the molecules are farther apart/ closer together in a liquid (so a gas is easy to compress/a liquid is difficult) *(1 mark)*

7. (a) (i) A *(1 mark)*
(ii) B and D *(1 mark)*
(iii) Hertz *(1 mark)*
(b) Sound *(1 mark)*
(to)
electrical *(1 mark)*

8. (a) Support of the body
Protection (of body organs)
Movement (together with muscles)
Production of blood cells
(2 marks for all four correct; 1 mark for three or two) *(2 marks)*
Helpful hint: *The first three marks should be straightforward to obtain, but the last point may have been more difficult to get.*

(b) The force/load/weight that is being moved/lifted *(1 mark)*
The distance moved by the load/weight *(1 mark)*
This distance of the muscle/effort from the pivot *(1 mark)*
Helpful hint: *Although this is a biology question, this question is based on moments, in physics. Be prepared to apply your knowledge and understanding from different topic areas.*

9. (a) Hydrogen *(1 mark)*
(b) (i) 1–14 *(1 mark)*
(ii) Red *(1 mark)*
(iii) Green *(1 mark)*
(c) Advantage: it is more sensitive/reads to one or more decimal places/Universal indicator (at best) reads to nearest pH *(1 mark)*
Disadvantage: it is not portable/a delicate instrument/needs calibrating/setting before use *(1 mark)*

10. (a)

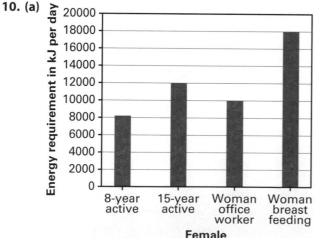

(1 mark for scale and labelled axes; 2 marks if all bars are drawn correctly; 1 mark if two or three bars are drawn correctly) *(3 marks)*
(b) To produce milk *(1 mark)*
(c) Larger energy requirement for male *(1 mark)*
(Males are generally) more muscular or have higher metabolic rate/any appropriate answer with a reason. (Allow 'Larger energy requirement from female' if an appropriate reason is given) *(1 mark)*

11. (a) 12% *(1 mark)*
(The gases listed add up to 88% (1 + 77 + 10)
∴ the percentage of other gases is 100 – 88 = 12)
(b)

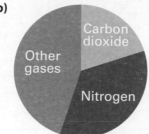

Correct calculations *(1 mark)*
Correct proportions in pie chart *(1 mark)*
Helpful hint: *You will be given credit for representing your proportions correctly, even if the calculations are not correct.*
To calculate each, it is a fraction of a full circle, i.e. 360°
e.g. for carbon dioxide, 20 ÷ 100 × 360 = 72°
You need to use a protractor to measure your angles
(c) There was not enough oxygen (for respiration) *(1 mark)*

12. (a) 10% *(1 mark)*
(b) (i) Double-glazing is expensive *(1 mark)*
Only 15% of the heat lost is through windows and doors *(1 mark)*
(ii) For 100-mm thick insulation – 12 years (300 ÷ 25) *(1 mark)*
For 270-mm thick insulation – 3 years (450 ÷ 150) *(1 mark)*
(c) Conduction *(1 mark)*
Convection *(1 mark)*

13. (a) Plants *(1 mark)*
(b)

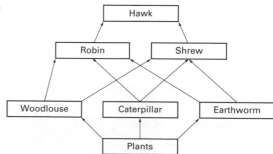

Shows producer at bottom *(1 mark)*
No or few errors in links *(1 mark)*
Arrows pointing in the correct direction *(1 mark)*
Helpful hint: *A common source of error among students is that the arrows are drawn in the wrong direction. The arrows show the direction the food is moving in.*

Set B, Test Paper 2

1. (a) Meter 1   Current
      Meter 2   Potential difference   *(1 mark)*
     Helpful hint: *Remember that ammeters are connected in series to the circuit; voltmeters in parallel.*

(b) 1.69 (ohms)   *(1 mark)*
     The calculation is $1.10 \div 0.65$

(c) Repeat the readings   *(1 mark)*

(d) (i) Alloy   *(1 mark)*

     (ii) Copper   *(0 mark)*
        Low electrical resistance/high conductivity is important for transmitting electricity
                              (1 mark)

        Helpful hint: *Note that there is no mark for answering copper; the mark is for your explanation.*

(e) (i) A long coil of wire   *(1 mark)*
     (ii)   *(3 marks)*

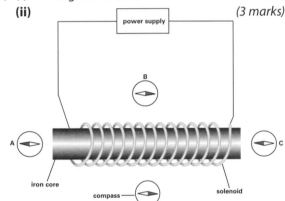

2. (a) (i) Phoebe   *(0 mark)*
        The tip of the plant is in her direct eye line (reducing error)   *(1 mark)*

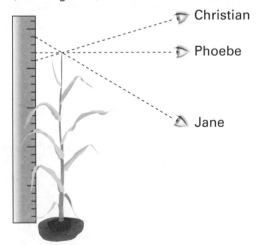

        Helpful hint: *Note that there is no mark for answering Phoebe; the mark is for your explanation.*

     (ii) Accurate   *(1 mark)*

     (iii) Systematic (error)   *(1 mark)*

(b) Palisade (cell)   *(1 mark)*
     Chloroplast   *(1 mark)*

(c) Stoma   *(1 mark)*
     It enables the plant to exchange gases between itself and the atmosphere   *(1 mark)*
     Helpful hint: *The examiner would accept an answer relating to taking in carbon dioxide for photosynthesis, but other gases are exchanged also (e.g. oxygen and water vapour).*

(d) Two answers from:
     Its flat structure/large surface area;
     air spaces in leaf for gases to move;
     large numbers of stomata to exchange gases
                              (1 mark)

3. (a) Hydrogen   *(1 mark)*

(b) (i) $100cm^3/30°C$, experiment 2   *(1 mark)*
     (ii) An outlier or an anomalous result   *(1 mark)*
     (iii) 40°C   *(1 mark)*

(c) (i)

Graph: y-axis "Average volume of gas produced in cm³" from 0 to 600; x-axis "Temperature in °C" from 0 to 60. Points plotted at approximately (10,100), (20,210), (30,310), (40,405), (50,510) with best fit line.

     2 marks for four or five points correct; 1 mark for two or three points correct   *(2 marks)*
     (ii) Best line (see above)   *(1 mark)*
     (iii) An increase in temperature results in an increase in the volume of gas produced
                              (1 mark)
        The volume approximately doubles with every 10°C rise   *(1 mark)*
        Helpful hint: *You should have found the first mark straightforward to obtain, but the second is more difficult. Look out for these types of relationships in your data.*

(d) Filter the liquid;   *(1 mark)*
     collect the filtrate   *(1 mark)*
     Helpful hint: *You could add more sulfuric acid until the iron fillings have dissolved, but as iron fillings are very impure, you would still need to filter the liquid*

4. (a) A hypothesis   *(1 mark)*

(b)

Depth of water	Average speed in cm per sec
1	**31**
2	**44**
3	**53**
4	**61**

     (2 marks for all four correct; 1 mark for two or three correct)   *(2 marks)*

(c) To reduce error/if just one measurement were taken, it might not be a typical result *(1 mark)*

(d) The results support his idea *(1 mark)*

Helpful hint: Never say that a hypothesis/idea has been proven. Evidence can support a hypothesis, but never prove it (even if there's a large amount of evidence in favour of it).

5. (a) (i) Solute *(1 mark)*

(ii) Solvent *(1 mark)*

(b) (i) 110g *(1 mark)*

(ii) The solubility of sodium chloride does not change/changes very little as temperature is increased *(1 mark)*

(iii) 22°C (allow 21–23°C) *(1 mark)*

Helpful hint: It is not possible to read accurately from this graph, so you will be allowed ±1°C for your answer.

(c) Sodium nitrate *(1 mark)*

6. (a)

Bowler	Bowling speed in m/s
Jimmy	40
Mitchell	42
Monty	24
Ryan	39
Stuart	41

(3 marks for all four correct; 2 marks for three, two or five correct; 1 mark for one correct) *(3 marks)*

(b) 97.2 kilometres per hour *(2 marks)*

Convert 27×3600 metres per hour

= 97 200km per hour *(1 mark)*

There are 1000 metres in 1km, so then divide this figure by 1000 *(1 mark)*

7. (a) (i) 100 times *(1 mark)*

The concentration is 4% in the air breathed out, and 0.04% in the air breathed in. So the calculation is $4 \div 0.04 = 100$.

(ii) Respiration *(1 mark)*

(iii) glucose + oxygen *(1 mark)*

$(\rightarrow)$

carbon dioxide + water *(1 mark)*

Helpful hint: The equation is the reverse of photosynthesis (in photosynthesis, light energy is required to drive the reaction; here, energy is released), but don't get the processes mixed up.

(b) Two from: The burning/combustion of fossil fuels (for energy); *(1 mark)*

increased burning of fossil fuels; increased energy consumption; increased number of cars; deforestation *(1 mark)*

8. (a) Pressure = force/area *(1 mark)*

(b) (i) 60 *(1 mark)*

N/cm² *(1 mark)*

Helpful hint: Substituting the values into the formula, the calculation is:

Pressure = $\frac{F}{a} = \frac{240}{4} = 60$

Note that one mark has been awarded for the units.

(ii) 960(N) *(1 mark)*

Helpful hint: Substituting the values into the formula, the calculation is:

Pressure = $\frac{F}{a}$

$60 = \frac{F}{16}$

$\therefore F = 960$

(c) Gases are compressible/liquids are not (to any extent) *(1 mark)*

Liquids are therefore better at transferring force into pressure *(1 mark)*

9. (a)

(3 marks for all four correct; 2 marks for three correct; 1 mark for one or two correct) *(3 marks)*

(b) From the mother/in the mother's blood; *(1 mark)*

through the placenta/along the umbilical cord; *(1 mark)*

by diffusion *(1 mark)*

(c) (i) Alcohol abuse has decreased from (around 45%) in 1992 to (around 20%) in 2007 *(1 mark)*

Cannabis abuse has increased from (around 19%) in 1992 to (around 46%) in 2007 *(1 mark)*

Helpful hint: When describing something, it helps to add numbers/values to your description if you can.

(ii) Answers related to impaired/reduced development of the foetus *(1 mark)*

10. (a) Can't be seen/invisible *(1 mark)*

(b) (i) The distance from the sunlamp to the plastic film should be the same *(1 mark)*

Use the same thickness/amount of sunscreen each time *(1 mark)*

Helpful hint: The examiner would also accept answers referring to using the same type of sunlamp or same type/thickness of plastic film.

(ii) Test a piece/pieces of the plastic film without a coating of sunscreen *(1 mark)*

(c) Heat zinc in air

or

Heat zinc carbonate/thermal decomposition of zinc carbonate *(1 mark)*

11. (a) Precise but not accurate *(1 mark)*

(b) Accurate and precise *(1 mark)*

(c) Neither precise nor accurate *(1 mark)*

1. (a) (i) A *(1 mark)*
       Helpful hint: *Students often confuse the cell membrane and cell wall. The cell membrane is pushed up against the cell wall.*
   (ii) B *(1 mark)*
   (b) **Two** answers from:
       (Its cells have) cell walls;
       chloroplasts;
       large vacuoles *(2 marks)*
       Helpful hint: *Animal cells often have vacuoles too, but they're never this large. They're also not permanent structures in animal cells, but don't write this, as you can't tell this from the diagram.*
   (c) The reservoir contains very large numbers of these green organisms/algae; *(1 mark)*
       because of eutrophication/large amounts of fertiliser in the reservoir *(1 mark)*

2. (a) (i) Orange *(1 mark)*
       Green *(1 mark)*
       Violet *(1 mark)*
       Helpful hint: *There are a number of mnemonics for remembering the colours of the spectrum, e.g. **R**ichard **O**f **Y**ork **G**ave **B**attle **I**n **V**ain, or **R**un **O**ff **Y**ou **G**irls – **B**oys **I**n **V**iew. But you can also work them out, as the primary colours – red, green and blue – overlap.*
   (ii) Dispersion *(1 mark)*
   (b) If another prism is placed after the first (that has produced the spectrum); *(1 mark)*
       the colours are recombined *(1 mark)*
       or
       If light from **one** of the colours of the spectrum from the first prism is passed through another prism; *(1 mark)*
       it will not produce a spectrum *(1 mark; maximum 2 marks)*

3. (a) Mixture *(1 mark)*
   (b) (i) Propane   **gas**
          Butane   **liquid** *(1 mark)*
       (ii) Propane   **liquid**
          Butane   **solid** *(1 mark)*
   (c) propane + oxygen *(1 mark)*
       ($\rightarrow$)
       carbon dioxide + water *(1 mark)*
   (d) **Two marks from:** Temperatures are colder in the winter/warmer in the summer *(1 mark)*, so (changing proportions) will make mixture more fluid (less viscous)/flow more easily in winter; *(1 mark)*
       and change how easily the LPG ignites *(1 mark)*

4. (a) (i) C *(1 mark)*
       (ii) A *(1 mark)*
       (iii) E *(1 mark)*

   (b) (i) Stomach *(1 mark)*
       (ii) Small intestine *(1 mark)*
   (c) Enzyme *(1 mark)*

5. (a) (i) 90 (J) *(1 mark)*
       (ii) Heat *(1 mark)*
       (iii) 10% *(1 mark)*
   (b) 40% *(1 mark)*
       Helpful hint: *The calculation is (600/1500 $\times$ 100)%.*
   (c) *(1 mark per row)* *(4 marks)*

Appliance	Power in W	Power in kW	Time appliance is used for in hours	Units of electricity used in kWh
Hair dryer	**2000**	2.0	1	**2**
Kettle	3000	**3.0**	2	**6**
Light bulb	100	**0.1**	20	**2**
Toaster	**1300**	1.3	1	**1.3**

   Helpful hint: *In questions where you have to fill in tables, you will often get one mark for each correct row.*

6. (a) (i) Triceps *(1 mark)*
       (ii) Contracts/shortens *(1 mark)*
       (iii) Muscles cannot get longer – they can only contract (or be relaxed); *(1 mark)*
           so they have to work in (antagonistic) pairs – when one contracts, the bone is moved one way, when the other contracts, the bone is moved the other way *(1 mark)*
   (b) (Fibre B, a ligament) is used to link a bone to another bone *(1 mark)*
       (Material C, cartilage) acts as a shock absorber between bones at joints *(1 mark)*

7. (a) (i) Carbon, hydrogen and oxygen *(1 mark)*
       Helpful hint: *Note that you need all three elements correct to get the mark. This is often the case in straightforward questions.*
       (ii) 23 *(1 mark)*
   (b) (i) *(1 mark)*

       (ii) *(1 mark)*

       Helpful hint: *Do not confuse the oxidising symbol with the flammable symbol.*
       (iii) Use in a well-ventilated space/do not breathe the vapour/keep away from flames when using/contact avoid with eyes and (prolonged) contact with skin *(1 mark)*

8. (a) (i) engine *(1 mark)*

(ii) Air resistance/drag

Friction (between tyres and road surface)

(2 marks)

(iii) (Forward and retarding) forces are balanced

(1 mark)

(b)

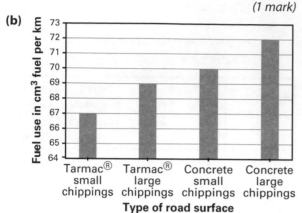

(2 marks for all bars drawn correctly; 1 mark for three or two bars drawn correctly) *(2 marks)*

9. (a) (i) Ceramic *(1 mark)*

(ii) Polymer *(1 mark)*

(b) A mixture; *(1 mark)*

of more than one metal or a metal and other elements *(1 mark)*

(c) From the solidification/slow crystallisation of magma/molten rock *(1 mark)*

10. (a) (i) The greater the number of turns, the stronger the electromagnet *(1 mark)*

(ii) Increase the current *(1 mark)*

(b) (i) The paperclips would fall off; *(1 mark)*

as a magnetic field is only produced when current flowing *(1 mark)*

(ii) Steel forms a permanent magnet/ stays magnetic when current is turned off. *(1 mark)*

Helpful hint: *Remember that electromagnets can be switched on and off by turning the electricity on and off. Permanent magnets are needed in electric motors, generators, microphones and speakers.*

11. (a)

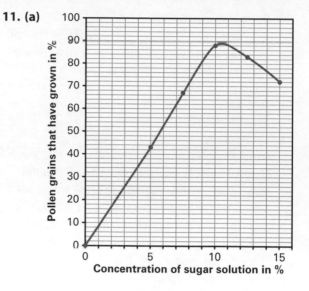

(2 marks for all six points plotted correctly; 1 mark for three, four or five points plotted correctly) *(2 marks)*

Appropriate line of best fit *(1 mark)*

Helpful hint: *Unless you're told to, don't just join the points. Look for a trend in the data and draw a line of best fit. Decide whether the line of best fit should be a straight line or a curve. Draw a line as close to the points as possible. Balance the number of points below and above the line.*

(b) (i) Percentage of pollen grains that have grown (after one hour) *(1 mark)*

Helpful hint: *Many students are confused as to which is the dependent variable. The dependent variable is usually what you measure in the experiment.*

(ii) Type of plant/spider flower or length of time/ one hour or number of pollen grains *(1 mark)*

(c) The percentage germination increases as the sugar concentration increases; *(1 mark)*

and reaches a maximum at around 10/11%; *(1 mark)*

then decreases *(1 mark)*

(d) One answer from: (Raw material for) growth **or** respiration/for energy *(1 mark)*

12. (a) Copper *(1 mark)*

(b) (i) Aluminium *(1 mark)*

(ii) Palladium *(1 mark)*

(c) (i) Gold is a very rare metal while aluminium is in plentiful supply *(1 mark)*

(ii) The world's natural sources of a metal are finite/ would one day be used up (even though they might be in plentiful supply now); *(1 mark)*

it reduces waste (that has to be disposed of somehow); *(1 mark)*

mining (for more metals) destroys the environment/landscape *(2 marks)*

Set C, Test Paper 2

1. (a) (i)   (They move) from a high concentration (on
               the cotton wool);   *(1 mark)*
               to a low concentration (in the tube)   *(1 mark)*
        (ii) (They move by) diffusion   *(1 mark)*
        (iii) Molecules   *(1 mark)*
    (b) (i)   A chemical reaction (between the ammonia
               and hydrogen chloride)   *(1 mark)*
        (ii) The ammonia diffuses more quickly/the
               hydrogen chloride moves more slowly *(1 mark)*
               Helpful hint: *This is because ammonia has
               the lower relative molecular mass (is 'lighter'),
               but if you give this answer, it doesn't quite
               answer the question.*

2. (a) Pressure = force/area
        The area is $(0.3 \times 0.3)m^2 = 0.09m^2$ *(1 mark)*
        ∴ Pressure = 72/0.09
                   $= 800N/m^2$ *(1 mark)*
        Helpful hint: *Show your working. One mark is
        awarded for this.*
    (b) (i)   Sedimentary   *(1 mark)*
        (ii) Formed when shells, sand and mud/layers/
               sediments are deposited at the bottom of
               water;   *(1 mark)*
               layers are compacted into rock   *(1 mark)*
    (c) (i)   Metamorphic   *(1 mark)*
        (ii) Formed when limestone is subjected to
               heat and/or pressure   *(1 mark)*

3. (a) The combination of the organisms and their
        environment that make up a particular
        habitat   *(1 mark)*
        Helpful hint: *In your answer, you must make sure
        that you mention* both *the organisms and the
        environment.*
    (b) carbon dioxide 1 water   *(1 mark)*
        (→)
        carbonic acid   *(1 mark)*
        Helpful hint: *In a chemical equation, examiners
        usually award 1 mark for getting the reactants
        correct, and one mark for the products.*
    (c) (i)   The burning/combustion of fossil fuels/
               trees/waste or volcanoes   *(1 mark)*
               Helpful hint: *Your answer must refer to
               burning/combustion, unless you have written
               'volcanoes'. The respiration of organisms also
               releases carbon dioxide, but you wouldn't
               usually refer to this release as an emission.
               Your answer would be accepted, however.*
        (ii) Slow decrease from 1850 to 1950   *(1 mark)*
               More rapid decrease from 1950 to the
               present day   *(1 mark)*
        (iii) −0.11   *(1 mark)*
        (iv) Continuation of line (extrapolation) shown
               appropriately   *(1 mark)*

2250 (or appropriate answer)   *(1 mark)*
Helpful hint: *You will be allowed some
tolerance with your answer. You may not have
found that it was the year 2250. You must
draw a line on the test paper from the current
gradient of the graph. You must read off
where it corresponds with a decrease of 0.5.*

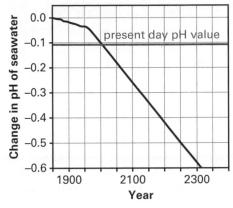

        (v)   That the decrease in pH will continue at the
               present rate   *(1 mark)*

4. (a) (i)   Iodine and astatine   *(1 mark)*
        (ii) Astatine   *(1 mark)*
               Helpful hint: *Some of the values are close
               together, so you need to be careful with your
               calculations. Jot them down as you do them.*
    (b) (i)   The melting point increases as you go down
               the group (from fluorine to astatine) **or**
               The boiling point increases as you go down
               the group (from fluorine to astatine) *(1 mark)*
        (ii) The reactivity decreases as you go down
               the group (from fluorine to astatine) **or**
               The reactivity increases as you go up the
               group (from astatine to fluorine)   *(1 mark)*
        (iii) Bromine and potassium chloride are
               produced   *(1 mark)*

5. (a) (i)   Genetic (variation)   *(1 mark)*
        (ii) Better camouflaged on the ice/in the
               snow;   *(1 mark)*
               so more able to catch prey   *(1 mark)*
    (b) Pale-coated bears pass on these genes;   *(1 mark)*
        so over time become paler and paler, and
        become a new species   *(1 mark)*

6. (a) (i)   28 days   *(1 mark)*
        (ii) By the Earth's gravity   *(1 mark)*
               Helpful hint: *To get the mark, you need to
               say it's the Earth's gravity.*
    (b) (i)   5   *(1 mark)*
        (ii) At phase 1   A   *(1 mark)*
               At phase 7   G   *(1 mark)*

7. (a) The number of bubbles of gas is affected by/
        depends on the amount of light/brightness of
        light/light intensity   *(1 mark)*

Helpful hint: A hypothesis is a suggestion a scientist makes to try to explain something they've observed. Scientists then do experiments to test the hypothesis to see if it can be supported by the data they collect.

(b) (i) (The sodium hydrogencarbonate) is a source of carbon dioxide *(1 mark)*
Helpful hint: You may know the answer because you've done this experiment in school. But if you haven't, the clue is also in the chemical formula.

(ii) Photosynthesis *(1 mark)*

(c) (i)

carbon dioxide + water $\xrightarrow{\text{light energy}}$ glucose + oxygen *(2 marks)*

(ii) $6CO_2 + 6H_2O \rightarrow C_6H_{12}O_6 + 6O_2$ *(2 marks)*

8. (a) Two from: Volume of water; volume of fuel; distance between the spirit burner and the boiling tube *(2 marks)*

(b) (i) Flammable: keep away from flames
Toxic: use in a fume cupboard/avoid breathing vapour/use in a well-ventilated area *(1 mark)*

(ii) The hot spirit burner, boiling tube or water/ use of the glassware or thermometer *(1 mark)*

(c) (i) 16.8; 19.0; 20.9 *(1 mark)*
Helpful hint: The results are recorded to one decimal place in each result, so you can't have more in the average. You could lose marks if you write down more.

(ii) Butanol *(1 mark)*

(d) Exothermic *(1 mark)*

9. (a) (i) 40 dB *(1 mark)*

(ii) 1 kHz *(1 mark)*
Helpful hint: Use a ruler to line up the points on the graph with the value on the y-axis, or better still, draw a horizontal line.

(b) (i)

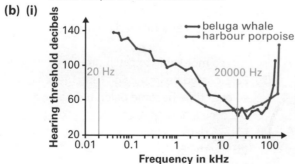

Both lines or points correct *(1 mark)*
Helpful hint: Be careful with units. 1 kHz = 1000 Hz. You need to make the conversion.

(ii) The hearing range shows some overlap with that of the beluga whale and harbour porpoise; *(1 mark)*

but the beluga whale and harbour porpoise can hear sounds of a higher pitch/ frequency; *(1 mark)*
while humans can hear sounds of a lower pitch than the harbour porpoise, and to some extent, the beluga whale *(1 mark)*
Helpful hint: Your answers might not quite correspond with those above. Like all answers in science, it will be judged on quality, and not quantity.

10. (a)

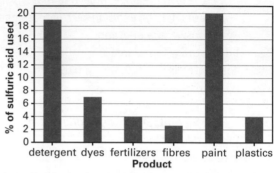

(2 marks for six correct bars; 1 mark for five, four or three) *(2 marks)*

(b) (i) Oxidation *(1 mark)*

(ii) The mass of the product is the same as the mass of the reactants. *(1 mark)*
One molecule of sulfur trioxide is formed from one molecule of sulfur dioxide. *(1 mark)*

(iii) $SO_3 + H_2O$ *(1 mark)*
$(\rightarrow)$
H_2SO_4 *(2 marks)*

11. (a) Two answers from:
Slow overall decrease in smoking from 1948 to 1970;
in two stages/with some small rises around 1956 and 1960;
then steady decrease between 1970 and 2010 *(2 marks)*

(b) (i) There has been a decrease in the rate of smoking and lung cancer *(1 mark)*
The decrease in lung cancer lags behind the fall in smoking *(1 mark)*

(ii) Lung cancer has other causes/not everyone who gets lung cancer smokes *(1 mark)*

12. (a) A substance that dissolves another substance *(1 mark)*

(b) Chromatography *(1 mark)*

(c) So that he could compare the distance travelled between the cocaine and the sample from the crime scene; *(1 mark)*
so as to identify any cocaine (in the sample from the crime scene) *(1 mark)*

(d) Contains cocaine; *(1 mark)*
and two (unknown) substances *(1 mark)*